A Global Access Strategy for the U.S. Air Force

David A. Shlapak, John Stillion, Olga Oliker, Tanya Charlick-Paley

Prepared for the United States Air Force
Approved for public release; distribution unlimited

Project AIR FORCE

The research reported here was sponsored by the United States Air Force under Contract F49642-01-C-0003. Further information may be obtained from the Strategic Planning Division, Directorate of Plans, Hq USAF.

Library of Congress Cataloging-in-Publication Data

A global access strategy for the U.S. Air Force / David A. Shlapak ... [et al.].
p. cm.
"MR-1216-AF."
Includes bibliographical references.
ISBN 0-8330-2959-2
1. United States. Air Force—Foreign service. 2. Air bases, American. I. Shlapak, David A.

UG634.49 .G57 2002
358.4'1357'0973—dc21

2001016077

RAND is a nonprofit institution that helps improve policy and decisionmaking through research and analysis. RAND® is a registered trademark. RAND's publications do not necessarily reflect the opinions or policies of its research sponsors.

Cover design by Maritta Tapanainen

Published 2002 by RAND
1700 Main Street, P.O. Box 2138, Santa Monica, CA 90407-2138
1200 South Hayes Street, Arlington, VA 22202-5050
201 North Craig Street, Suite 202, Pittsburgh, PA 15213-1516
RAND URL: http://www.rand.org/
To order RAND documents or to obtain additional information, contact Distribution Services: Telephone: (310) 451-7002;
Fax: (310) 451-6915; Email: order@rand.org

PREFACE

This report represents the culmination of a concerted effort within RAND's Project AIR FORCE to examine the political, operational, logistical, and force protection issues associated with overseas basing for the Expeditionary Aerospace Force. The result of this effort, presented here, is a strategy for global access and basing of U.S. aerospace forces. This study builds on a body of previous RAND research relating to enhancing the United States Air Force's expeditionary capabilities, including:

- Paul S. Killingsworth, Lionel Galway, Eiichi Kamiya, Brian Nichiporuk, Timothy L. Ramey, Robert S. Tripp, and James C. Wendt, *Flexbasing: Achieving Global Presence for Expeditionary Aerospace Forces*, MR-1113-AF, 2000
- Robert S. Tripp. Lionel Galway, Paul S. Killingsworth, Eric Peltz, Timothy L. Ramey and John G. Drew, *Supporting Expeditionary Aerospace Forces: An Integrated Strategic Agile Combat Support Planning Framework*, MR-1056-AF, 1999
- Lionel Galway, Robert S. Tripp, Timothy L. Ramey, and John G. Drew, *Supporting Expeditionary Aerospace Forces: New Agile Combat Support Postures*, MR-1075-AF, 2000
- John Stillion and David T. Orletsky, *Airbase Vulnerability to Conventional Cruise-Missile and Ballistic-Missile Attacks: Technology, Scenarios, and U.S. Air Force Responses*, MR-1028-AF, 1999.

This work was sponsored by the Deputy Chief of Staff for Air and Space Operations (AF/XO) within the Air Staff and should be of interest to planners and operators within the Air Force. It may be of value to policymakers elsewhere in the Department of Defense and the U.S. government who are involved in arranging and maintaining relationships that can either facilitate or hinder other states' cooperation with the United States in the full range of military operations.

Our research was conducted within the Strategy and Doctrine program of Project AIR FORCE. Comments are welcome and should be directed to the Program Director, Edward Harshberger, or to the lead author, David Shlapak (David_Shlapak@rand.org).

Primary research for this study concluded in late 1999, so events of 2000 and 2001 are not fully reflected here. However, nothing that has transpired would, in our opinion, dramatically alter our conclusions. Indeed, the terrorist attacks on the United States in September 2001 and the subsequent "war on terrorism" reinforce our main premise, which is that the United States—and the Air Force—must prepare for challenging contingencies in unexpected places at inconvenient times.

PROJECT AIR FORCE

Project AIR FORCE, a division of RAND, is the Air Force federally funded research and development center (FFRDC) for studies and analysis. It provides the Air Force with independent analyses of policy alternatives affecting the development, employment, combat readiness, and support of current and future aerospace forces. Research is performed in four programs: Aerospace Force Development; Manpower, Personnel, and Training; Resource Management; and Strategy and Doctrine.

CONTENTS

FIGURES

TABLES

SUMMARY

The United States Air Force (USAF) has undertaken a number of initiatives aimed at improving its responsiveness and effectiveness in fast-moving, quickly evolving contingencies. Whether confronting a humanitarian crisis in Africa, sustaining a peacekeeping operation in Southwest Asia, or fighting a major war in Korea, the USAF has sought to increase its contributions to deterrence, crisis response, and war fighting when called on to respond to challenges to U.S. interests.

To accomplish this goal, the Air Force has instituted significant changes in its organization, operations, doctrine, and planning. Having reconstituted itself as an "expeditionary aerospace force," or EAF, the Air Force is now in the process of changing many aspects of how it does business. This report is intended to contribute to this process by helping the Air Force think through one critical aspect of its future: access for basing.

Many important components of U.S. power projection capabilities, including land-based fighters and Army divisions, rely on access to overseas installations, foreign territory, and foreign airspace. The Army has no role other than homeland defense if its forces do not venture outside U.S. borders, and the Marine Corps' *raison d'etre* is the conduct of expeditionary operations "from the halls of Montezuma to the shores of Tripoli." Even the Navy's carrier battle groups, free of the need for foreign bases per se, nevertheless require access to foreign ports and facilities for resupply and other support functions.

Similarly, access and basing issues are of great importance to the USAF. Like the Army, USAF forces are for the most part equipped and configured to fight from "in theater"; fighters and attack aircraft such as the A-10, F-15, F-16, and F-117 have unrefueled combat radii of 300–500 nm. And while these ranges can be greatly extended through aerial refueling, such aircraft cannot be used to best advantage when they are based thousands of miles from their intended targets.[1] Moreover, Air Force operations have experienced real difficulties because of access problems, most recently during a series of post-*Desert Storm* crises in the Gulf from 1996 to 1998. Earlier, the emergency airlift to Israel during the 1973 Middle East war and the 1986 *El Dorado Canyon* punitive strike on Libya were similarly hampered by access difficulties.

For a variety of reasons, these issues are not expected to disappear in the coming years:

- First, despite many predictions that the nation-state will become increasingly irrelevant, we see no evidence that governments are losing control of their physical territory.[2]
- Second, the kinds of contingencies that crop up in the next decade or two will likely occur in areas where the United States faces sizable access uncertainties.
- Finally, compounding the problem is the fact that evolving threats may induce planners to consider basing air forces farther away from enemy territory.

The USAF thus faces a complicated set of demands as it confronts its future as an expeditionary force. It must plan, organize, equip, and train itself according to a new set of principles suited to a world that

---

[1]Chapter Three includes a detailed analysis of the reasonable limits of extended-range operations for most current USAF fighters and attack aircraft. We also note that the next planned generation of USAF tactical aircraft, the F-22 and the F-35 Joint Strike Fighter (JSF), do not feature increased range among the advantages they will have over existing platforms.

[2]There are several examples of failed or weak states such as Somalia whose central control of their territory is uncertain at best. However, even in these cases, *someone*—a local warlord perhaps, or rebel faction—exerts *de facto* authority over the real estate in question, and U.S. military operations in, from, or above that territory must take into account the desires of the controlling authority, whether legitimate or not.

demands frequent, short-notice deployment and employment across a spectrum of conflict that may occur virtually anywhere in the world. Moreover, the Air Force must do so in the face of grave uncertainties—driven by ineluctable political and military realities—regarding where, how, and when it will be able to operate. The USAF therefore needs a *global access and basing strategy* that will help it prepare for tomorrow's requirements. This report outlines an approach to such a strategy and recommends some specific components thereof.

THE POLITICS OF ACCESS

Understanding how circumstances have affected other countries' past decisions about U.S. access can help the Air Force better prepare for contingencies to come. We reviewed the history of U.S. access around the world, region by region, to draw out lessons that might help planners and others lay a firmer groundwork for ensuring adequate future access. This analysis led us to a set of implications for the Air Force.

First, we identified six factors that seem to have a profound effect on other countries' decisions regarding whether to cooperate with the United States in a given situation. Three factors that seem to favor cooperation are

- Close alignment and sustained military connections,
- Shared interests and objectives, and
- Hopes for closer ties with the United States.

Three factors that appear to work against cooperation are

- Fear of reprisals,
- Conflicting goals and interests, and
- Domestic public opinion.

Our global survey suggests that two fundamental tools available to the United States are particularly appropriate to helping ensure access. The first such tool—*transparency and information sharing*—can help convince friends and allies that their interests do not in fact conflict with those of the United States and that cooperation with the United States aligns with their own goals. The second tool, *engagement*—which is directed mainly at states where ties are less clear and less strong—helps establish that the United States is a good friend to have in one's corner and thus someone for whom doing an occasional favor may be wise. Maintaining an active program of military-to-military contacts and using U.S. "information dominance" to help shape the perceptions of partner countries and other aspects of engagement may be the best assurance that, when the need arises, U.S. military forces can find adequate access to perform their missions both quickly and safely.

This analysis also suggests that access is likely to prove most troublesome in two regions that are critical to U.S. national security: the Persian Gulf and Asia outside the immediate vicinity of the Korean peninsula. In addition, sub-Saharan Africa and Latin America—particularly in the far south—will pose serious operational challenges. In these areas and perhaps elsewhere as well, situations will almost certainly arise in which USAF forces will confront missions that must be undertaken with less-than-optimal access and basing.

OPERATIONAL CONSIDERATIONS AFFECTING ACCESS REQUIREMENTS

With this as our background, we set out to evaluate how less-than-optimal access—by which we mean, in essence, basing farther away from the target area than is standard USAF practice—would affect the operational capabilities of a forward-deployed USAF force. Toward that goal, we explored air expeditionary task force (AETF) operations in a notional scenario involving an attack by Iran on Kuwait and Saudi Arabia. In this analysis, we

- Identified potential basing options for both the fighter and support elements of the AETF
- Selected alternative pairs of beddown locations (one base for fighters and another for support assets) to study the impact of increased distance between bases and targets
- Employed a sortie-generation model to estimate the AETF's combat capability from each set of bases

- Adjusted key parameters determining operational effectiveness and repeated the process.

Our work focuses on the effects of being forced to base at more distant locations owing to enemy offensive capabilities that seriously threaten closer-in facilities. However, the effects we identify and the remedies we recommend would be equally applicable to a situation in which political constraints or lack of usable close-in infrastructure limits basing options. This analysis suggested the following conclusions:

- Based strictly on aircraft operating characteristics, a number of locations are suitable for AETF deployment in Southwest Asia. However, geography, political factors, adversary threat capability, and commanders' willingness to accept risk could interact to limit such choices, especially for large, vulnerable support aircraft.
- This narrowing of options could lead AETFs to deploy to fields far from their intended targets, requiring long missions that are hard on fighter crews and that consume large quantities of fuel.
- The combat capabilities of an AETF can decrease dramatically when the aircraft are forced to base at increasing distances from their intended operational areas.
- In the short run, modest increases in fighter crew ratios and tanker support could allow the typical AETF to operate with about the same effectiveness from ranges of 1000–1500 nm to target as a "nominal" AETF can from about 500 nm.[3]

ACCESS IN OPERATIONS OTHER THAN WAR

Many—indeed, most—future overseas military actions will be of the kind often referred to as "military operations other than war" (MOOTW). Although such actions have been steadily increasing in frequency, the Department of Defense has been inclined to view MOOTW as lesser-included cases for force planning and basing ar-

---

[3]At longer ranges, it may also be desirable to replace the A-10s in a standard AETF with faster jets, such as F-16s.

rangements. However, one can conceive of a plausible scenario in which the United States is involved in—and the Air Force is supporting—a *major* peacekeeping and humanitarian mission in a remote area with relatively little intelligence or logistical support and with limited infrastructure. Without adequate planning, this type of mission could present a daunting challenge in terms of both rapid deployment and manageable sustainment. Keying off recent experiences in Somalia (1992–1993) and Rwanda (1994), we created just such a scenario for a peacekeeping and humanitarian crisis centered in Burundi. Our exploration of this scenario, coupled with our analysis of past experience, led us to believe that

- These complex MOOTW could impose significant demands on Air Force lift and logistical capabilities.
- The desire for a rapid response to a quickly deteriorating situation could be frustrated by a lack of adequate prior planning and coordination as well as by a dearth of infrastructure to support a major airlift.
- Demands on specialized USAF units, such as engineers, security forces, and aerial port squadrons, could be high in a challenging MOOTW.

In sum, our work suggests that future complex MOOTW could prove highly demanding for the USAF and probably should not be dismissed as lesser-included contingencies. Instead, more planning may be called for to ensure that the Air Force is both operationally and politically prepared to manage such missions.

DEVELOPING A GLOBAL ACCESS STRATEGY

Access will, in short, remain a challenge both to the U.S. military in general and to the Air Force in particular for the foreseeable future. On the positive side

- The United States enjoys strong defense relationships with a large and growing number of countries around the world. This web of engagement serves to facilitate access for the USAF.

- While access has historically proven to be an irritant on many occasions, U.S. diplomacy, flexibility, and luck have usually resulted in the availability of workarounds to enable operations.
- There are a number of countries that, in looking to improve or cement their security relations with the United States and the West, could be strong candidates for enhanced access arrangements.
- Given some modifications in manning and support, current and future USAF forces appear capable of sustaining a reasonably high tempo of operations (OPTEMPO) at fairly long ranges (up to 1000–1500 nm) from their operational areas.

The negatives are as follows:

- "Assured access" outside U.S. territory is a chimera. National sovereignty may be eroding in cyberspace, but in the "real world" of air bases and airspace, it continues to reign supreme.
- Even close allies, such as the UK and Germany, have at times refused the United States access or overflight.
- In addition to the politically driven access problems that the United States has occasionally encountered, new military threats—particularly advanced surface-to-surface missiles—may change the calculus of risk, inducing commanders to base forces farther away from the immediate combat zone.
- Access arrangements in Southwest Asia and Asia outside of Korea and Japan are limited and may prove woefully inadequate for the kinds of contingencies that could develop in those regions.
- Given current and likely future access arrangements, it could prove difficult to project and sustain a significant amount of power into sub-Saharan Africa and Latin America south of the equator. The former appears to present particularly serious challenges.

In short, the USAF confronts a complex set of circumstances with regard to access and basing. What options exist for dealing successfully with them?

We have identified five approaches for managing access and basing in the future. They are

- Expand the number of overseas main operating bases (MOBs) to increase the likelihood that forces will be present where and when they are needed.
- Identify one or more "reliable" allies in each region of the world and count on them to cooperate when asked.
- Proliferate security agreements and alliances to broaden the set of potential partners in any given contingency.
- Negotiate and secure long-term extraterritorial access to bases, as was done with Diego Garcia.
- Rely on extended-range operations from U.S. territory.

We believe that each of these strategies is insufficient in and of itself to ensure adequate access. We therefore recommend that a hybrid strategy be adopted to deal with future demands on the USAF. We further suggest that the USAF consider a metaphor from the financial world and treat the construction of an appropriate access and basing strategy as a problem in *portfolio management*. We consider this analogy to be sound along several lines:

- As on Wall Street, the environment USAF planners face is one dominated by *uncertainty*. In such a "market," a well-hedged portfolio is the best path to success.
- Managing risk and exploiting opportunity require *diversification*. Success will depend on having a range of contingency options, plans, and capabilities.
- *Information flows* are critical to good decisionmaking. The United States must be aware of its partners' sometimes divergent goals, strategies, and interests. *Engagement* and *transparency* play pivotal roles.

What sort of portfolio might the USAF seek to construct? In keeping with the metaphor, we will describe one possibility in terms of three components: core investments, hedges against risk, and opportunities to watch out for.

We suggest three core investments:

- The United States should maintain its current array of overseas MOBs in Europe and Asia.
- The USAF should establish a small number of *forward support locations* (FSLs) worldwide. Essentially a "mega-MOB" intended to support power projection, the United States would preposition spares, equipment, and munitions at these locations. Five FSLs—in Alaska, Puerto Rico, and the UK and on Guam and Diego Garcia—could put most of the earth's surface within C-130 range of a major USAF hub.
- The United States should seek to maintain and expand its contacts with key security partners worldwide. Although there would appear to be no need to pursue additional formal defense ties as a means of shoring up prospects for access, consistent engagement is of great value.

In terms of hedging against risk, we have two principal suggestions:

- First, both planning and force packaging may need to become more responsive to possible access constraints. Otherwise, basing and access limitations could impose significant penalties on expeditionary operations.
- The USAF should also consider ways of extending the reach of its combat air forces, by either developing a fast, longer-range strike platform or deploying a new generation of long-range munitions for carriage by existing and planned strike aircraft.

Finally, we recommend that the USAF explore two avenues for exploiting potentially lucrative opportunities. First, it should conduct preliminary analyses to determine the feasibility of "renting a rock"—i.e., establishing a sovereign U.S. presence along the lines of Guam or Diego Garcia on some uninhabited atoll or islet—in the Western Pacific. Second, given the rapid pace of geopolitical change over the past ten years, the USAF should take careful note of as-yet unappreciated opportunities to engage new partners as possible access sites and should pursue those that seem most valuable.

As a final piece to the portfolio puzzle, we would like to highlight two regions where we believe current access arrangements to be insufficient and where the risk of being called to action is in our view high. Both immediate and longer-term ameliorative steps may be needed to shore up the USAF position in Southwest Asia (where the problem is driven by the seeming impossibility of gaining firm commitments from America's regional friends) and in much of Asia (where geography and politics conspire to create difficulties).[4]

In the near term, we believe that flexible planning will be critical to ensuring the USAF's ability to effectively fly and fight in the Persian Gulf. Enabling deploying forces to maintain OPTEMPO from nonoptimal basing locations could be important in this region. Looking further, broadening the list of possible strategic partners is advisable as well, with Israel being a prime candidate should a broad peace accord allow for its "normalization" in the region.

At the same time, the current USAF basing posture along the Pacific Rim is inadequate to support high-intensity combat operations anywhere much beyond the Korean peninsula. Especially problematic is the lack of bases available in the vicinity of the Taiwan Strait. Renewed access to bases in the northern Philippines could be immensely helpful here, especially if confidence were high that such bases could be used were a fight to erupt between Mainland China and Taiwan. Such political concerns—which are rife with regard to Taiwan throughout the region—would make "rent-a-rock" a particularly attractive option here.

Still farther south, the United States may want to consider taking steps to improve its access prospects by increasing the level and extent of its presence in Singapore. Malaysia also appears interested in improved relations with the United States, and this may create an opportunity to increase USAF access there. Thailand and Vietnam are candidates as well.

In the longer term, an increased number of longer-range combat platforms (or short-legged platforms with long-range munitions) would prove useful in both the Gulf and East Asia.

Our research indicates that there is no panacea or "silver bullet" waiting to be discovered with regard to access and basing. Old

---

[4]Some Asian basing issues are discussed in Khalilzad et al. (2001).

problems such as the vagaries of international politics will persist, and new ones—dozens or even hundreds of long-range accurate missiles aimed at U.S. bases—doubtless will emerge. Furthermore, nothing comes free: There are real costs, in terms of both money and opportunity, associated with any course of action the USAF might take to deal with potential problems in this area. This is the bad news.

On the other hand, we do not emerge from our work with merely a tale of woe. To the contrary, we believe that the problems that exist are manageable and that even those that can not be foreseen—always the most worrisome—can be minimized by a well-thought-out global access strategy. The strategy we suggest calls for increased flexibility and pays off in enhanced robustness against the unavoidable uncertainty that characterizes this problem. In the final analysis, then, access is not a problem to be solved—it is a portfolio to be managed.

ACKNOWLEDGMENTS

This study benefited from the efforts of numerous people to whom the authors are indebted. James C. Wendt played an important role in helping shape our thinking about the political dimensions of overseas access. Maren Leed contributed to our early brainstorming about the problem, and Jennifer Kawata assisted in synthesizing our initial findings. We thank them all.

Paul Killingsworth and Richard Moore provided incisive reviews of an earlier version of this report; what you now hold in your hands is much improved for their influence. Current and past RAND colleagues Daniel Byman, Peter Chalk, Roy Gates, Zalmay Khalilzad, Jerry McGinn, Bruce Pirnie, Daniel Raymer, William Stanley, and Alan Vick all contributed their time and insights to our cause. Major Mike Pietrucha and Dr. Robert Mullins of Headquarters, U.S. Air Force Directorate for Strategy, Concepts and Doctrine (AF/XOXS), provided comments and assistance that proved invaluable.

Dionne Sanders, Lisa Rogers, and Amy Sistek provided able and reliable administrative support during the writing and publication of this book.

Finally, we extend many thanks to our editor, Jeanne Heller, for her patience, good humor, and expertise, in equal measures.

To all of these people, and the others whose contributions we have inadvertently overlooked, we give our thanks; what is useful in our work owes much to them. We wish we could also share the blame with them for any errors or shortcomings in what follows, but, alas, for those we alone must be held responsible.

ACRONYMS

ACRI	African Crisis Response Initiative
AEF	Aerospace Expeditionary Force
AETF	Air expeditionary task force
AMC	Air Mobility Command
APOD	Aerial port of debarkation
AWACS	Airborne Warning and Control System
CALCM	Conventional air-launched cruise missile
C^2	Command and control
C^4ISR	Command, control, communications, computers, intelligence, surveillance, and reconnaissance
CAS	Close air support
CENTRASBAT	Central Asian Peacekeeping Battalion
CONUS	Continental United States
CRAF	Civil Reserve Air Fleet
CSAR	Combat search and rescue
CVBG	Carrier battle group
DRC	Democratic Republic of the Congo

EAF	Expeditionary Aerospace Force
ECOMOG	West African Peacekeeping Force
FSL	Forward support location
GPS	Global Positioning System
HARM	High-Speed Anti-Radiation Missile
HD/LD	High demand/low density
HDR	Humanitarian daily ration
HTS	HARM Targeting System
IDP	Internally displaced person
JASSM	Joint air-to-surface standoff missile
JDAM	Joint Direct Attack Munition
JSF	Joint Strike Fighter
JSOW	Joint standoff weapon
JSTARS	Joint Surveillance and Target Attack Reconnaissance System
LANTIRN	Low-Altitude Targeting Infrared System for Night
LCN	Land capacity number
LD/HD	Low-density/high-demand
MOB	Main operating base
MOG	Maximum on ground
MOOTW	Military operation other than war
MP	Military police
MTW	Major theater war
NEO	Noncombatant evacuation operations

NGO	Nongovernmental organization
OAS	Organization of African States
OPTEMPO	Operational tempo
PAX	Passengers
PCN	Pavement classification number
PfP	Partnership for Peace
PGM	Precision-guided missile
PRC	People's Republic of China
SAC	Strategic Air Command
SAM	Surface-to-air missile
SDB	Small diameter bomb
SEAD	Suppression of enemy air defense
SF	Security forces
SLOC	Sea-lanes of communication
SSB	Small smart bomb
SWA	Southwest Asia
TALCE	Tanker airlift control element
TDY	Temporary duty
TFW	Tactical Fighter Wing
TOLD	Takeoff and landing data
UAE	United Arab Emirates
UNHCR	United Nations High Commissioner for Refugees
UNSCOM	United Nations Special Commission

USAID	United States Agency for International Development
WMD	Weapons of mass destruction

Chapter One

INTRODUCTION

It has been said that Great Britain conquered its earth-girdling empire at least in part to provide coaling stations to support the Royal Navy's global mastery of the high seas. The United States in the 21st century will similarly require robust and flexible basing and access for its wide-ranging aerial fleets. With imperial conquest out of fashion, however some other strategy must be devised to ensure that the U.S. Air Force (USAF) and its sister services have access to the bases and facilities they need for rapid and effective operations.[1]

The USAF has undertaken a number of initiatives aimed at improving its responsiveness and effectiveness in fast-moving, quickly evolving contingencies. Whether confronting a humanitarian crisis in Africa, sustaining a peacekeeping operation in Southwest Asia (SWA), or fighting a major war in Korea, the USAF has sought to increase its contributions to deterrence, crisis response, and war fighting when called on to respond to challenges to U.S. interests.

To accomplish this goal, the Air Force has instituted significant changes in its organization, operations, doctrine, and planning.

---

[1]The word "access" can have multiple meanings. In current USAF vernacular, it is often used to denote not merely the basing or overflight rights needed to support operations but also the ability to gain dominance over adversary threats and thereby achieve some degree of tactical freedom—gaining "access" to the battlespace, as it were. In this usage, the "access issue" can include such elements as defeating an opponent's air-to-air and surface-to-air capabilities. This report takes the narrower view of the topic, addressing access in terms of threats—political and operational—to USAF *basing* for future contingencies.

For perspectives on the broader access issue, see Hawley et al. (2000); Wolfe (2001); Fulghum (2001); and Tirpak (2001).

Having reconstituted itself as an "Expeditionary Aerospace Force," or EAF, the USAF is now in the process of changing many aspects of how it does business. This report is intended to contribute to this process by helping the Air Force think through one critical aspect of its future: access for basing.

THE EXPEDITIONARY IMPERATIVE

For most of its history, the USAF has relied heavily on *forward basing*, maintaining a substantial portion of its "tactical" force[2] structure at overseas bases from which they would fight in the event of a war.[3] This was an appropriate strategy during the Cold War, when U.S. defense planning focused on deterring or defeating a Warsaw Pact attack on Western Europe. Even in the late 1970s and early 1980s, when concern over Western energy security brought the Persian Gulf into focus as a new area of critical concern, worries centered on a Soviet attack into Iran as a prelude to the "real" war, which would be fought on the plains of Europe. USAF fighter squadrons were spread across the map of NATO, with wings and squadrons based at various times in France, Iceland, Italy, the Netherlands, Spain, Turkey, the UK, and West Germany. In the Pacific, Air Force units were located on Guam and in Japan, the Philippines, and South Korea.

With the 1990s came the Warsaw Pact's implosion and the collapse of the USSR, removing the impetus for a massive U.S. presence in Europe. Bases closed and units came home, many to be disestablished as the USAF force structure drew down. By and large, the remaining forces were centrally based in the United States, deploying to available overseas bases only if circumstances so required.[4]

---

[2]"Tactical" here means those forces that are not primarily or exclusively committed to the nuclear retaliatory mission, performed until the early 1990s by the USAF's Strategic Air Command (SAC). It is worth noting that until the parallel deployment of the B-52 and KC-135 tanker in the mid-to-late 1950s, SAC, too, depended on overseas basing for its mainstay force of B-47 medium bombers.

[3]Despite the technical inaccuracy of the practice, this report will use "overseas" as a synonym for "outside the territory of the 50 United States."

[4]The differences in the posture of the USAF's tactical forces before and after the Cold War are difficult to overstate. In 1982, 30 USAF fighter and tactical reconnaissance squadrons were permanently based at four locations in the UK, one in Spain, one in the Netherlands, and five in West Germany. In late 1999, only nine squadrons re-

In the midst of this evolution, Iraq invaded Kuwait and Operation *Desert Shield* was launched in response to that aggression. Although the United States had no permanent main operating bases (MOBs) on the Arabian peninsula, it benefited from Saddam Hussein's decision to sit tight after overrunning Kuwait. Thus, nothing interfered with the five-month-long buildup of Coalition forces in Saudi Arabia and elsewhere in the region, a buildup that was greatly facilitated by the wealth of bases and infrastructure available on the Arabian peninsula as well as years of prior cooperation between the United States and regional militaries. Indeed, by the time *Desert Storm* began on January 17, 1991, most USAF units were flying from bases that, in terms of logistics and operational support, were nearly as well endowed as full-fledged MOBs. Only after the hugely successful Gulf War did cracks begin to show in the USAF's planned post–Cold War posture. Two main factors contributed to the stresses that made themselves felt at this time.

First, while U.S. forces remained in the Gulf region to help enforce various UN resolutions binding Iraq, U.S. partners in the region remained reluctant to permit the United States to establish permanent bases on their territory. Saudi Arabia, the possessor of the area's most extensive and robust base infrastructure, proved particularly shy in this regard. As a result of this lack of cooperation on the parts of key friends and allies, the USAF has been forced to rely on a series of temporary deployments to carry out its part in Operations *Northern* and *Southern Watch*. By the mid to late 1990s, these seemingly interminable activities were taking their toll on readiness—costs that have been further exacerbated by a steady stream of additional overseas commitments, the second necrotic element.

Indeed, life after the "end of history" has proven to be quite busy for the USAF. While operations above Iraq constituted a steady drain on Air Force resources, other contingencies—such as famine relief in Somalia, peace enforcement in Bosnia, and something approaching an air-only major theater war (MTW) over Kosovo and Serbia—imposed surge demands that have at times stressed USAF resources to the limit. Since late 2001, the USAF has also undertaken Operations

mained home-based in Europe, at Lakenheath in Britain, Aviano in Italy, and Spangdahlem in Germany.

Noble Eagle and *Enduring Freedom*, adding dramatically to the drain on USAF resources.[5]

To cope with these pressures, the Air Force leadership decided that major changes were needed in the service's organization and the focus of the USAF was therefore shifted from reliance on forward-based forces to rapidly responding to dynamic situations. Thus was born the idea of an air (later "aerospace") expeditionary force (AEF)—a task-organized unit that could quickly deploy to a trouble spot and begin sustained operations within 48 hours of being ordered out of garrison.[6]

As the AEF concept was articulated and elaborated, it became clear that this concept might also hold the key to addressing the USAF's ongoing dilemma with sustained temporary duty (TDY) deployments. Building from the idea of the AEF, the Air Force reconceptualized itself as an EAF. At the heart of the EAF concept are ten permanent AEFs, each having some 134 aircraft available from designated squadrons, groups, and wings, which rotate through a 15-month schedule during which each AEF has a 90-day period of susceptibility for overseas deployment.[7] The USAF hopes that by providing some predictability to the prospect of TDY deployments, the EAF structure will mitigate the stress on service members and facilitate smoother responses to the "steady state" demands on its forces.[8]

---

[5]*Noble Eagle* is the homeland air defense operation begun in the aftermath of the September 11 terrorist attacks; *Enduring Freedom* is the military operation against the Al Qaeda terrorist organization and the former Taliban regime in Afghanistan.

[6]Current USAF Chief of Staff General John Jumper is usually regarded as the author of the AEF concept, which was developed when he was commander of the U.S. Ninth Air Force responsible for USAF operations in the Gulf. The 48-hour timeline may prove extremely challenging in situations where forces are deploying with little notice to an area that is not rich with prepositioned stockpiles of equipment and munitions. See Galway et al. (1999).

[7]These AEFs, which are organizational constructs, are different from the "AEFs" that began carrying out short-term deployments to SWA in the mid-1990s; this is one example of how the terminology surrounding the AEF/EAF has not always been as clear as could be hoped.

[8]The AEF is primarily a force management tool designed to help regularize the demands on USAF personnel to support day-to-day requirements for rotational presence overseas. Hence, its deployment rotations resemble the deployment schedule the Navy maintains for its fleet of flattops. And, just as carrier battle groups (CVBGs) can be unexpectedly called from port or out of workups to respond to an emerging crisis—with concomitant disruption to the orderly cycle of CVBG activity—so can USAF units

The AEF and EAF have thus emerged as a response to the tension between a post–Cold War reduction in permanent forward-basing opportunities and a rising demand for short-term foreign operations. The same tension has raised questions of access and basing to a new level of visibility. If the Air Force is not based on foreign soil but is still expected to operate overseas with little notice, it needs a strategy that will maximize its chances of gaining adequate access to perform its missions effectively and safely. This report attempts to shed some light on the nature and specifics of one possible approach to this problem.

The work reported here builds on a body of prior RAND research aimed at enabling effective USAF expeditionary operations. Although our problem is defined differently and our approach is our own, it should not be surprising if some of our conclusions echo those of the earlier efforts. In particular, we owe an intellectual debt to the team led by Paul Killingsworth, whose two-year study of AEF operations served as a foundation for our work.[9] As will be seen, our analysis reinforces virtually all of that team's key conclusions, especially their emphasis on the need for flexible planning frameworks and basing arrangements.

THE CHALLENGE OF ACCESS

Many important components of U.S. power projection capabilities—such as land-based fighters and Army divisions—are highly reliant on access to overseas installations, foreign territory, and foreign airspace. The Army has no role other than homeland defense if its forces do not venture outside U.S. borders, and the Marine Corps' whole *raison d'être* is the conduct of expeditionary operations "from the halls of Montezuma to the shores of Tripoli." Even the Navy's carrier battle groups, free of the need for foreign bases per se, still require access to foreign ports and facilities for resupply and other support functions.

---

be called on to go into action outside their normal time "in the box" with similar impact on the scheduled rotation.

[9]Their work is documented in Killingsworth et al. (2000).

Access and basing issues are also of great salience to the USAF. Like the Army, the USAF's forces are for the most part equipped and configured to fight from "in theater," as evidenced by the fact that fighters and attack aircraft such as the A-10, F-15, F-16, and F-117 have unrefueled combat radii of 300–500 nm. And while such ranges can be greatly extended through aerial refueling, these aircraft cannot be used to best advantage when they are based thousands of miles from their intended targets.[10] Moreover, Air Force operations have experienced real difficulties because of access problems, most recently during a series of post–*Desert Storm* crises in the Gulf from 1996 to 1998.[11]

In September 1996, Iraq perpetrated a gross violation of the terms of the Gulf War cease-fire, launching a ground attack against Kurds in and around the northern Iraqi town of Irbil. The United States wanted to engage the attacking Iraqi forces using aircraft based in Turkey and Saudi Arabia—aircraft already flying missions over Iraq enforcing the northern and southern no-fly zones. However, both Ankara and Riyadh denied the United States the use of these aircraft for combat missions against the Iraqi troops. In addition, Jordan denied the United States the use of its airspace despite the fact that a USAF air expeditionary task force (AETF) had recently been deployed there. Deprived of the use of its land-based airpower, the United States launched cruise missile strikes against air defense and command-and-control (C^2) facilities in southern Iraq. These attacks had no obvious impact on the Iraqi army's operations against the Kurds and must generally be assessed as a failure.

Similar events have been repeated since that time:

- In November 1997, Iraq expelled six U.S. members of the United Nations Special Commission (UNSCOM) weapon inspection team. In response, the United States sent additional aircraft to the region and increased aerial reconnaissance over Iraq. Saudi

[10]Chapter Three analyzes the reasonable limits of extended-range operations for most current USAF fighters and attack aircraft. It is also worth noting that the next planned generation of USAF tactical aircraft, the F-22 and the F-35 Joint Strike Fighter (JSF), do not feature increased range among the advantages they will have over existing platforms.

[11]The following discussion draws heavily on unpublished work by James C. Wendt.

Arabia denied the United States permission to launch attacks from its bases and did not allow any additional forces into the country. Turkey was not asked for permission to conduct strikes from its territory but made it clear that if asked, it would refuse.

- Just two months later, the unresolved crisis flared again when Saddam Hussein blocked weapon inspectors from inspecting presidential palaces and other "sensitive sites." Under the weight of extraordinary U.S. arm-twisting, Kuwait and Bahrain gave assurances of cooperation in military operations.[12] Even under pressure, Saudi Arabia declined to support strikes on Iraq; Riyadh not only denied the use of U.S. aircraft based in Saudi Arabia but would not allow those aircraft to be moved to neighboring countries to conduct attacks from there. Faced with such unequivocal Saudi opposition, first Bahrain and then Kuwait backed away from their initial support of the United States. Qatar and the United Arab Emirates (UAE) also refused to allow the use of their territory, and Jordan, Turkey, and Egypt expressed opposition to any U.S. air strikes.
- Another incident occurred in November 1998, when Iraq announced an end to cooperation with UNSCOM inspectors. Although many Arab governments were markedly more critical of Iraqi actions than was previously the case, such governments remained unsupportive of U.S. military action against Baghdad. Most prominently, Saudi Arabia again refused the United States access to its facilities for offensive operations.

The USAF has also seen its activities impeded rather than stopped outright by access difficulties. Three examples illustrate this.

In 1973, President Nixon ordered an emergency airlift to resupply Israel, which had been attacked on two fronts by Arab armies and was fighting for its life. Operation *Nickel Grass*, as the airlift was named, was severely hampered by a lack of cooperation from America's European allies, which refused to permit USAF airlifters to transit their airspace or use their facilities while en route to or from Israel. Heavy pressure from the Nixon administration finally per-

---

[12]First, Secretary of State Madeleine Albright visited the region, followed closely by Secretary of Defense William Cohen.

suaded Portugal in essence to look the other way while U.S. C-5s and C-141s landed at and took off from Lajes airfield in the Azores. Absent this grudging assistance, the airlift—which Egypt's president Anwar Sadat later cited as one of the pivotal elements in his decision to request a cease-fire—would almost certainly have been impossible.[13]

Almost 13 years later, lack of support from NATO allies again complicated a U.S. military operation. In April 1986, President Reagan ordered air strikes on a number of targets in Libya in retaliation for alleged terrorist activities. Operation *El Dorado Canyon* was complex enough to begin with, involving as it did F-111 and EF-111 aircraft flying from Great Britain and U.S. Navy jets operating from two carriers in the Mediterranean Sea. These problems were multiplied, however when both Spain and France refused to allow the F-111s to fly over their territory during the mission. This resulted in a substantial lengthening of the flying times for the F-111s, which had to start the trip to their targets in the southeast first by flying southwest over international waters opposite the French and Spanish coastline and then by slipping through Gibraltar and across the Mediterranean (see Figure 1.1). Having followed this tortuous course on their inbound journey, the crews were then expected to avoid strong Libyan air defenses, deliver their weapons (subject to extremely stringent rules of engagement), and turn around and make their way back the way they came.

This prolonged trip necessarily took a toll on both men and machines. By the time the F-111s made it to Libya, numerous aircraft had had difficulties with their sensitive targeting systems that either prevented them from dropping the bombs they had carried such a distance or resulted in the delivery of the weapons well off target. Tired aircrew also made errors that resulted in improperly aimed ordnance. Thus, while on a strategic level the attack can arguably be

---

[13]In 1973, the USAF's fleet of C-141A transport aircraft was not fitted for aerial refueling and could not have flown nonstop from the U.S. East Coast to Israel. The C-5A, which was equipped for refueling but was prohibited from doing so because of difficulties with its wing structure, could have made the trip on one tank of gas, but its maximum payload would have been reduced to 33 tons. By stopping at Lajes, the C-5s were instead able to carry an average of 68 tons per sortie. See Lund (1990), and Comptroller General (1975), pp. 10, 30.

RAND MR1216-1.1

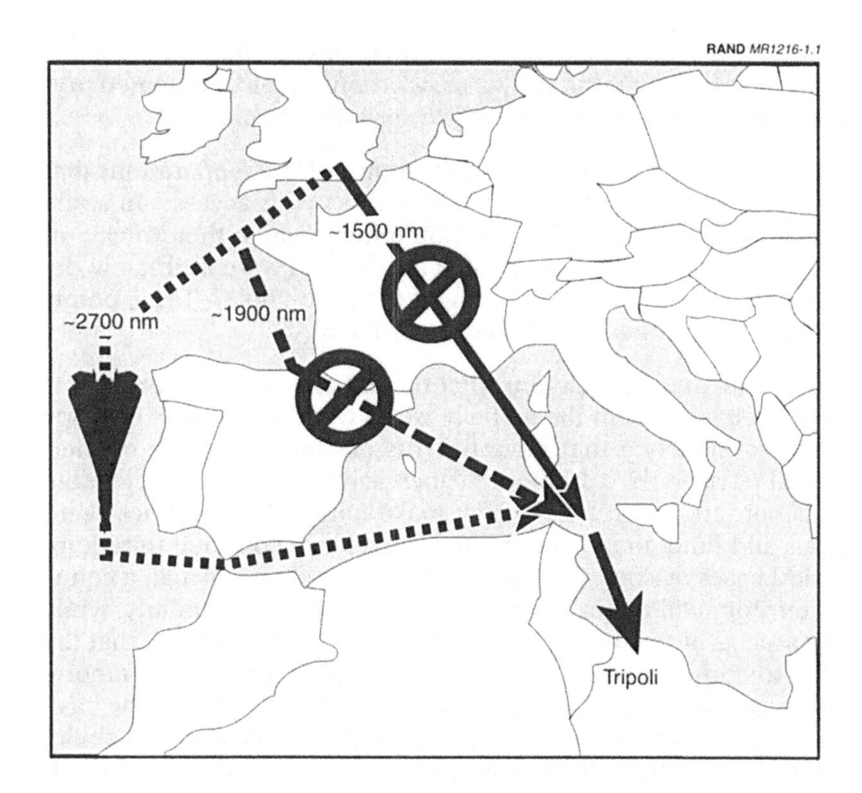

Figure 1.1—Schematic Mission Profile for Operation *El Dorado Canyon*

assessed as a success, tactically the strikes achieved significantly less than planners had hoped. At least some of the blame for the disappointing performance must be assigned to the excruciating mission profile, which stressed aircrew and aircraft well past the bounds of their normal operations.

Finally, in December 1998 UNSCOM reported that Iraq had not complied with UN demands that Baghdad dismantle its programs for developing and producing weapons of mass destruction (WMD). Acting to carry out earlier threats, U.S. and British air forces attacked Iraqi military forces, installations, and facilities suspected of being related to WMD. The United States was able to use bases in Kuwait and Oman to launch some strikes. However, both Saudi Arabia and Turkey—where the United States had its largest concentrations of deployed assets—denied the use of their bases. Operation *Desert Fox*, as the campaign was known as, was consequently executed predominantly by cruise missiles and carrier-based aircraft.[14]

The past thus contains numerous examples of USAF operations that have been adversely affected by difficulties with access. In some cases, these problems sufficed to stop things dead in their tracks; on other occasions, workarounds of various kinds were ultimately devised. What sorts of challenges might the future hold? Three points seem worth making to help frame the problem.

First, despite many predictions that the nation-state will become increasingly irrelevant in the globally wired world of the new century, we see no evidence that governments are losing control of their physical territory.[15] Although "cybercash" may flow unregulated across borders and refugees may make national boundaries seem porous and fluid, organized military forces will continue to require physical bases of operation, and their uninvited presence in a country's territory will retain its traditional significance. Similarly, while some aspects of sovereignty may well wither away, we expect that the ability to control access to bases and airspace will not be among those factors that diminish in importance. As it has been in the past, so in the future the idea of "assured access"—the guaranteed ability for the United States to do what it wants when it wants, where it wants, from and via foreign territory—will remain a chimera. Except in the most extraordinary circumstances, nations simply do not cede so much control over such fundamental things.[16] After all, having at

---

[14]Access difficulties have continued to be bothersome. During NATO's Operation *Allied Force* air campaign against Serbia, alliance member France reportedly refused to allow armed bombers flying from Fairford in the UK to overfly its territory enroute to their Balkan targets. Fulghum and Wall (2001).

[15]There are multiple examples of failed or weak states such as Somalia whose central control of their territory is uncertain at best. However, even in these cases, *someone*—a local warlord, perhaps, or rebel faction—exerts *de facto* authority over the real estate in question, and U.S. military operations in, from, or above that territory must take the desires of the controlling authority, legitimate or not, into account.

[16]The United States gained extraterritorial control over the Canal Zone in Panama by dint of good old-fashioned imperialism: physically occupying the real estate and refusing to give it back. A 1966 bilateral treaty gave the United States access for defense purposes to the British Indian Ocean Territories, including Diego Garcia, over which fly both U.S. and British flags.

least some control over acts of war committed from one's territory must be considered one of the defining qualities of a government.

Second, many of the contingencies that crop up in the next decade or two are likely to occur in areas where the United States faces sizable access uncertainties. Europe—where the United States enjoys a history of close security relationships, an enduring alliance superstructure, and a plethora of potential basing options—may continue to witness limited conflicts on its southern and eastern fringes. However, the probable foci for large-scale warfare lie in regions of problematic access: Southwest Asia, the Taiwan Strait and South China Sea, and South Asia all loom large as possible hot spots.[17] Africa, too, may be staring down the barrel of a series of humanitarian crises that will make past horrors pale in comparison as AIDS, ethnic rivalry, and still-exploding populations create seismic pressures that weak or corrupt governments will be unable to contain. Massive humanitarian intervention or peace operations in sub-Saharan Africa could present U.S. planners with particularly serious access problems.[18]

Finally, evolving threats may induce planners to reassess the calculus of access. Historically, the USAF has preferred to deploy its fighter forces to locations lying within easy reach of their intended operational areas, generally within a few hundred miles. There are, of course, good reasons for this preference: shorter missions mean higher sortie rates and maximum efficiency from a force of a given size. If adversaries have the capability to credibly threaten the security of these close-in bases using surface-to-surface missiles, Special Forces, or other means, future theater commanders may face difficult tradeoffs between bedding forces down either optimally or securely.[19] Under circumstances such as those depicted in Figure 1.2—a conflict with an Iran equipped with a number of *Nodong*-class

---

[17]The emergence of Central Asia as an unexpected theater of operations in late 2001 is evidence of the unpredictability of future requirements.

[18]See Chapter Four for an in-depth discussion of how basing and access issues might play out in a complex military operation other than war.

[19]Effective defenses against one or more of these various threats may be feasible in the future; to the extent that they are, these problems could be mitigated somewhat. The authors, however, are convinced that no plausible near- to midterm defenses will be so robust as to eliminate this risk-versus-efficiency calculus. This factor, combined

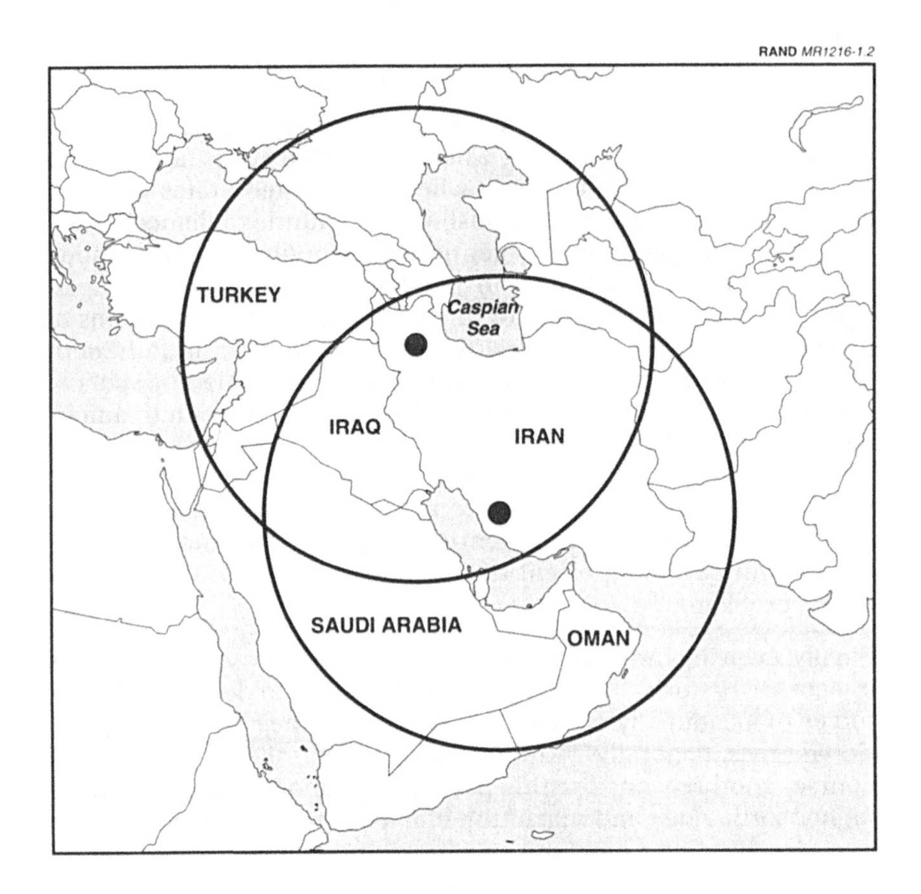

Figure 1.2—Range Rings for 700-nm-Range Missiles Based in Iran

700-nm-range missiles, potentially carrying WMD warheads—the USAF may want or need to fight from distant bases to improve prospects for force protection.[20] Plans and platforms must therefore be able to operate effectively under such suboptimal conditions.

---

with the ever-present risk of access denial by foreign governments, demands that the USAF develop the requisite operational flexibility to cope with the possibility of operating from longer ranges.

[20]The USAF has in the past operated from bases within the range of enemy ballistic missiles, notably during the 1991 Gulf War, and it could always choose to continue to

The Air Force thus faces a complicated set of demands as it confronts its future as an expeditionary force. It must plan, organize, equip, and train itself according to a new set of principles suited to a world that demands frequent, short-notice deployment and employment across a spectrum of conflicts that may occur virtually anywhere in the world. Moreover, it must do so in the face of grave uncertainties—driven by ineluctable political and military realities—with regard to where, how, and when it will be able to operate. The USAF therefore needs a *global access and basing strategy* that will help it prepare for tomorrow's requirements. This report outlines an approach to such a strategy and recommends some specific components thereof.

ORGANIZATION OF THIS REPORT

The remainder of this report consists of five chapters.

- Chapter Two reviews the region-by-region history of USAF access, with emphasis on patterns and trends that can help inform thinking on future opportunities and constraints.
- In Chapter Three, we employ quantitative analysis to help identify options for improving USAF operational capabilities in situations where forces are compelled—by friendly politics or enemy action—to conduct combat operations from distant bases.
- Chapter Four describes the demands that could arise in a complex military operation other than war (MOOTW) using a challenging peace-enforcement and humanitarian mission in Central Africa as an example.
- Chapter Five outlines our recommendations for a USAF global access strategy that is built around the idea of *portfolio management*. It also contains some brief concluding remarks.

---

do so. However, as threats increase in both quantity and quality, that choice may become riskier.

Chapter Two

THE POLITICS OF ACCESS

That adversaries will seek to prevent U.S. action by denying the United States access to territory and airspace is to be expected. That geography and nature itself will sometimes pose constraints, from mountain ranges to bad weather, is a fact of life. But what options exist when friends, allies, or neutral states deny the United States the use of their facilities and airspace or even of U.S. assets located on their soil? Insofar as the United States must respect such states' sovereignty over their own territory, these entities can prevent U.S. actions without using violence or force simply by saying "no." Unlike adversary-imposed constraints such as the destruction of base facilities or simple physical constraints such as distance, these *diplomatic* constraints on access present both advantages and disadvantages for military planners. While far more difficult to predict and hence to plan for, diplomatic constraints are at least somewhat susceptible to diplomatic counterefforts. Unlike an adversary or Mother Nature, states may be convinced to reverse their opposition to U.S. operations, thereby alleviating the problem.

One might expect that the network of relationships that exists between the United States and its friends and allies worldwide would help limit the number of occasions in which diplomatic access constraints emerge. The U.S. military has an excellent record of cooperation with a great many countries—a record that has facilitated U.S. access abroad for a wide range of activities. Yet there is tremendous variability to these relationships, and the links between ties and access can often prove tenuous indeed. Some of the United States' closest friends and allies have, for example, denied the United States the right to overfly their airspace for certain operations, as did Greece with regard to NATO combat aircraft during Operation *Allied Force*. Other friends, however, have proven more open to persuasion. Hungary, for example, allowed NATO combat forces to base on its soil for that same operation despite strong internal reservations about the impact cooperation with NATO might have on the ethnic Hungarian population in Serb-controlled Vojvodina.

Understanding how circumstances have affected other countries' decisions about U.S. access in the past can help the USAF better prepare for contingencies to come. In this chapter, we will discuss the history of U.S. access around the world, region by region, to draw lessons that will help planners and others lay a firmer groundwork for ensuring adequate access in the future.

THREE KINDS OF ACCESS

Permanent Presence

The presence of U.S. forces abroad, in bases or facilities that are operated by the United States either alone or in concert with host countries, constitutes an important kind of access. Today, the United States has substantial base presence (often referred to as "permanent" presence) in several NATO countries as well as in Japan, in Korea, and at Guantanamo Bay in Cuba.

With the exception of the Guantanamo Bay facility, all such permanent basing is hosted by allies that the United States is committed by treaty to defend. In fact, these garrisons—which have often served as focal points for U.S. military operations overseas—were established in large part to better equip the United States to effectively defend the countries in question. However, host-nation approval for use of these bases and facilities in missions not directly related to their intended purpose—defense of the host's territory—is by no means assured.

In Europe, the threat that for many years justified the ongoing U.S. presence there has largely evaporated. The risks of war in Korea, on the other hand, are such that allied support for the United States in the event of a conflict is virtually assured regardless of other disagreements with Seoul and Tokyo. In other situations, however, host countries may have little incentive to support U.S. actions that might conflict with their own interests. NATO ally Turkey, for example, did not allow the use of U.S. forces stationed at Incirlik to counter Iraqi intervention in the Kurdish civil war in 1996. Moreover, there are concerns that U.S. forces in Japan might not be permitted to participate should the United States decide to actively support Taiwan in a struggle with Mainland China.[1]

Mission Presence

In addition to permanent forward presence, the United States maintains substantial "mission" presence in countries where there is an ongoing military mission but to which there may or may not be a treaty commitment. The presumption in many of these contexts is that when the mission is over, U.S. troops will leave, as is the case with the current deployments in Saudi Arabia supporting Operation *Southern Watch*. Also in the mission-presence category are smaller deployments—such as the continuing naval and air support activities in Singapore—that lack the breadth and capability to qualify as true forward presence but that nonetheless contribute to the overall U.S. posture abroad. Missions of this sort may include defense of the host country and its interests, as in Kuwait and Oman, or may simply serve mutual needs, as in Australia. As with forward presence, however, having troops in place is no guarantee of the U.S. right to use them however and whenever it wishes. As noted previously, for example, Saudi Arabia has repeatedly prevented planned U.S. air strikes on Iraq when it has not shared the U.S. view of the necessity for strikes.

Limited Access

Finally, there are those countries where the United States maintains no forces on a regular basis but where its troops visit on occasion to assist in training, for exercises, or to take part in contingency operations. On each such occasion, of course, U.S. presence is subject to the invitation and/or approval of the host.

---

[1]The latter concerns may be ameliorated somewhat by the 1997 revision of the Guidelines for U.S.-Japan Defense Cooperation. See Ministry of Foreign Affairs of Japan, "The Guideline on Japan," Vol. 1, "Defense Cooperation," www.mofa.go.jp/region/n-america/us/security/guideline2.html, browsed 9 May 2002.

When it comes to employing U.S. forces in actual operations, existing arrangements for limited access can be helpful but, like permanent and mission presence, is hardly definitive. On the one hand, the physical presence of U.S. forces in place may make it easier for nations hosting ongoing U.S. deployments to permit use of their bases and facilities for contingency operations. However, many countries may for internal political and cultural reasons be sensitive to the long-term presence of foreign troops on their soil and attempts to negotiate ongoing access with these partners may thus be counterproductive. On the other hand, leveraging limited-access arrangements with such countries can help secure additional access when needed.

Formal Agreements and the Determinants of Access

Within all three of these access categories there is substantial variation in the extent to which U.S. presence is governed by formal agreements or arrangements. In some cases, explicit provisions exist governing access to the country and its facilities. In other instances, there may be an agreement regarding the legal status of U.S. troops in the country but little more. With some countries, including Saudi Arabia and several of the Partnership for Peace (PfP) states, no agreements of any sort exist, and issues are handled on a case-by-case basis.

In fact, neither the extent of U.S. presence nor the formalization of access arrangements appears to be a decisive factor governing whether a country will grant access to the United States in a given situation. To the extent that these arrangements and levels of U.S. presence reflect shared security needs, access will almost certainly be granted if both states feel it is necessary to meet those needs. Beyond that, however, no such guarantees exist.

Students of alliance behavior will not find this surprising. After all, it has been estimated that countries join their allies in war only about one-quarter of the time.[2] Although this estimate may not be heartening, these odds are significantly higher than those of *non*allies fighting alongside one another. Thus, it does not seem unreasonable

---

[2]Smith (1996), p. 17; see also Siverson and King (1980).

to assume that while presence and formal commitments may not guarantee access, they are likely to improve the chances that access will be granted. The historical record appears to support this assumption.

Another contributing factor that is potentially even more important than presence—because presence, after all, is limited to only a few countries—may be political-military ties and relations. Close military-to-military ties by and large suggest at least some shared security interests and are thus potentially indicative of a proclivity to cooperate in pursuit of common goals. U.S. military ties with other countries are diverse, ranging from the mutual defense commitments noted above to programs of contacts and exercises that may or may not be backed by formal agreements. Such formal agreements also vary, comprising those that regulate military assistance; those that formalize access arrangements, as discussed above; and agreements and arrangements regarding contacts, arms sales, and the like. Not all states with which the United States has contacts have arrangements that spell out such agreements; in some cases, these are states with which ties are comparatively close—as witness, once again, Saudi Arabia.

The U.S. experience with the combination of presence, ties, and access varies from region to region, as each part of the world presents both different access needs and different political and diplomatic environments. It is therefore crucial to consider each such region separately, both to draw appropriate lessons from the past and to better define future needs. Conclusions about access that can be derived from these regional analyses can to some extent be generalized, and similar patterns are evident across the regions. However, as far as actual basing and operations are concerned, it behooves us to identify issues that are region-specific, so that strategies can be devised to address each.

REGIONAL PERSPECTIVES AND PROSPECTS

Europe

In many ways, Europe is the United States' access gateway to much of the rest of the world. The United States has relied on its substantial forward presence in Europe not only for local missions, but also for operations in the Middle East and Africa. Moreover, while the need has not yet emerged for such a contingency, Europe could readily be considered a potential base for efforts in South Asia as well.[3] Europe's rich infrastructure, modern economies, and strong historical ties to the United States have made it an obvious choice to support and facilitate a wide range of combat, peacekeeping, and humanitarian operations—a situation that can be expected to continue. An obvious example was *Desert Storm*, in which an entire corps of U.S. Army forces stationed in Europe was moved to Saudi Arabia.

In addition to the large number of forces on the ground, U.S. military ties with European states not only are substantial but have grown over the past decade. Half a century of security commitment to NATO has now been expanded to embrace three new NATO members. Moreover, the NATO PfP initiative, which the United States sponsored, has increased the cooperation sphere to include 17 additional European countries and several Central Asian states. While there is no security commitment on the part of the United States to the non-NATO PfP states, there are substantial and growing programs of military contacts with several of them. Further, the desire on the part of several of these countries to achieve full-fledged NATO membership may affect their willingness to support U.S. efforts both in Europe and worldwide.

However, even close friends can disagree. Thus, while overall support has been excellent and relations good, a few outstanding cases point up the kinds of problems that can emerge.

The 1973 airlift to Israel and the Operation *El Dorado Canyon* strikes on Libya have already been mentioned in this context. In the former case, the United Kingdom, Spain, Italy, Greece, and Turkey all refused to provide any support to the United States or even to allow U.S. aircraft to overfly their territories,[4] while Spain, France, Germany, and Italy turned down requests to support the Libyan

---

[3]Indeed, European bases have been deeply involved in supporting the counterterrorist campaign in Afghanistan.

[4]Boyne (1998); Comptroller General of the United States (1975); Lund (1990); and Timsar (1981).

raid.[5] Then, three years after *El Dorado Canyon*, Spain asked that the U.S. 401st Tactical Fighter Wing (TFW), based at Torrejon Air Base, leave the country. The public call for the F-16s' departure was an outgrowth of a 1986 promise by Spanish Prime Minister Felipe Gonzalez that U.S. presence would be reduced in the face of rising anti-American sentiment in the country. Yet another contributing factor may well have been the participation in the Libyan strike of two KC-10 refueling tankers flying out of Zaragoza without authorization and, according to the Spanish government, without Madrid's knowledge. While U.S. access to other bases in the country was retained, the United States acceded to Spain's request, and the fighter wing was moved to Italy.[6]

Most recent U.S. military operations in Europe have focused on the former Yugoslavia, where both the Bosnia and Kosovo actions have included U.S./NATO air strikes. Although these operations were authorized and conducted by NATO, Greece refused to allow the alliance's combat forces to fly over its territory or to use its bases, although it did provide logistical support and allow humanitarian overflight.[7] Tellingly, Greece's behavior must be contrasted with that of Albania and Bulgaria, neither of which is a NATO member, and with that of Hungary, which became a NATO member after the Bosnia operation but before Kosovo. All three countries permitted overflight, and Hungary and Albania also hosted U.S. and NATO forces on their soil. Furthermore, Bulgaria did so despite facing a similar domestic situation to that in Greece, with significant ethnocultural linkages to the Serbs fueling high levels of public opposition to the bombings. Hungary, in turn, had to overcome substantial domestic concern that its support of NATO actions might endanger the large ethnic Hungarian community in Vojvodina, a region in Serbia.[8]

---

[5]Boyne (1999); Stanik (1996); Doerner (1986); Hersh (1987); "Allies Wanted 'All-out' Attack" (1986); Church (1986); Owen and Brown (1986).

[6]Schumacher (1986); Steele (1987); Riding (1990); Aguirre (1988); Mann (1988); "U.S., Spain Announce Withdrawal of U.S. F-16s" (1988); Cody (1987).

[7]"The First 8 Days" (1999); Abdallah (1999); "Orthodox but Unorthodox" (1999); "Air Ban on Turkish Fighter Planes" (1999); "Stifling U.S. Pressure" (1999).

[8]Tagliabue (1999); "Brave Gamble" (1999); "Balkan States Back NATO" (1999); Jordan (1999a); Jordan (1999b); Szamado (1999); Fitchett (1999); "NATO Deployed" (1999); and Sly (1999). While the governments in question may have differed in the degree to

European access problems have thus been a constraint on U.S. freedom of action. These experiences show that participation in NATO and even a history of the closest possible ties with the United States, as with the United Kingdom, do not ensure that access will be granted. What, then, drives such unwillingness to cooperate?

Fear of reprisal, be it economic or terrorist, was the most common reason European allies cited for their failure to support U.S. policy both in 1973 and in 1986. In fact, Portugal did suffer repercussions for its support of the United States in 1973, enduring a complete cutoff of oil supplies from the Arab states.[9] In 1986, European leaders questioned not only whether the planned U.S. action would invite reprisal but also whether it would be particularly effective in curtailing terrorism. Later reports suggested that evidence linking Libya to the terrorist attack in a Berlin discotheque that served as a spur to the air strikes may not have been entirely convincing to foreign governments.[10] On the other hand, the British prime minister's justification to Parliament of her decision to grant the United States access was based in part on shared U.S. intelligence.[11]

The more recent Greek situation is somewhat different and considerably less straightforward. What is particularly telling is that NATO member Greece gave in to whatever combination of public opposition and traditional tension with Turkey existed there while other allies and even non-NATO states cooperated with the United States despite what appeared to be equally valid reservations.

Clearly, Greece was willing to take the risk of angering the United States and NATO in refusing to go along with the rest of the alliance. However, it seems clear that Athens understood that anger and dissatisfaction were probably all that it risked, as there was no danger that the alliance would turn its back on this long-time NATO member. Hungary, by contrast, as a new NATO member, feared precisely such a rejection, remaining uncertain as to the solidity of the al-

---

which they supported NATO's goals in Kosovo, their dramatically different responses to the crisis—despite many similarities in their situations—would appear to require a more complex explanation.

[9]Timsar (1981).

[10]Hersh (1987).

[11]Owen and Brown (1986).

liance's commitment to its defense and security. Diplomatic efforts on the part of other NATO states and of the alliance as a whole were thus successful in convincing Hungary to open up its territory to NATO aircraft for the Kosovo mission. Bulgaria and Romania, which hope to be invited to join NATO, and Croatia, which has yet to be asked to join the PfP, were even more inclined to respond affirmatively to NATO pressure. The situation was similar in 1973. Portugal, a NATO member, was globally isolated and mired in a colonial war in Africa. Looking for support wherever it might be found, Lisbon was more susceptible to pressure from the United States than were other NATO countries, whose international positions were stronger.[12]

The *Allied Force* experience may thus be viewed as featuring several Portugals and, for the short term at least, may be a valuable lesson for European policy. The greater security concerns and perceived dependence on the United States of new NATO members and NATO aspirants may well render such states more cooperative and more susceptible to persuasion than long-term NATO allies such as Greece or France. Supporting this thesis is the fact that PfP partner Ukraine has offered NATO territory for use as a training range, which NATO has accepted.[13] Recently, Azerbaijan even raised the possibility of establishing bases on its territory—an offer that has been received with considerable ambivalence by the United States.[14]

The extent to which the United States can take advantage of such opportunities depends in part on its ever-evolving relationship with Russia. While ties with the post-Soviet states continue to be built and strengthened, the realization that Russia continues, to varying degrees, to see those states as lying within its sphere of influence has limited the West's willingness to reach out to them. Russia's own cooperation in European security since the end of the Cold War has been variable, reflecting its unique interests and concerns. The game of leverage that would be required to make full use of the entire post-Soviet space is thus sufficiently complex that, insofar as other options remain available, too strong a reliance on the post-Soviet states is unlikely and probably inadvisable.

---

[12]Timsar (1981).

[13]"U.S. Defense Secretary, Ukrainian Leaders Discuss" (1999).

[14]"Baku Asks for US Support" (1999); "Foreign Minister Zulfugarov Says" (1999).

In summary, the situation in Europe continues to be broadly favorable, if only because the options there are so plentiful and diverse that occasional setbacks are fairly easy to overcome. This does not mean, however, that new options should not be pursued, as the experience of being but one country away from mission failure, as occurred with *Nickel Grass,* is always a possibility. Focusing attention on and building ties with PfP states and Russia may provide just the opportunity the United States will need at some point in the future, while also potentially enhancing its overall reach further east.

Southwest Asia and the Middle East

As already noted, bases and forces in Europe have repeatedly been used to support U.S. operations in the Middle East and North Africa. Turkey, a NATO member that straddles the two regions, has been particularly integral to activities in both regions. But the United States has not relied solely on Europe for its Middle East operations. While U.S. permanent presence as defined above is maintained only in Turkey, the United States has maintained sizable forces in Saudi Arabia since the end of the Gulf War in 1991. While Turkey is the only state in the area to which the United States has a formal security commitment, U.S. airmen, soldiers, and sailors also operate in Egypt, Saudi Arabia, Oman, the UAE, Qatar, Kuwait, and Bahrain, supporting a range of missions that include but are not limited to those related to Gulf security.

Access in the Gulf region has always been limited and case-specific. Before the Gulf War, the U.S. *modus operandi* in the region was to come in, do what it planned to do, and leave. This pattern was determined not by U.S. preferences—indeed, Washington continually pressed for improved and formalized access arrangements—but rather by the refusal of friends, particularly Saudi Arabia, to support a more permanent presence. In fact, of all the Gulf states, Oman is the only one with which the United States has formal access arrangements that predate *Desert Storm.*

The Saudis have, however, repeatedly granted the United States contingency access. In 1987, when the Iran-Iraq War spilled over into attacks on Saudi and Kuwaiti shipping, the Saudis supported and facilitated Operation *Earnest Will*, the U.S. response of reflagging and escorting Kuwaiti ships.[15] In 1990, heavy pressure from the United States—coupled with American intelligence sharing that convinced Riyadh of the Iraqi threat to the kingdom—induced the Saudis to permit an enormous deployment of men and equipment to their country.[16] Following the Gulf War, Riyadh broke with tradition by allowing the United States to maintain some presence, as did a number of other countries in the region. However, while formal arrangements for access and defense ties have since been negotiated with Kuwait, Bahrain, and Qatar, Saudi Arabia remains a notable standout, refusing to formalize the relationship in any meaningful and enduring way.

Further, as with the Europeans, the granting of base and facility access to U.S. forces has not guaranteed *carte blanche* for their use. While no-fly zones continue to be enforced by U.S. and British forces based in Saudi Arabia and Turkey, the use of these bases for additional missions was denied on multiple occasions in 1996, 1997, and 1998.[17] Furthermore, persistent Saudi reluctance to allow aircraft based on its territory to engage in punitive strikes against Iraq continues to hinder operations.[18]

In some aspects, then, the Gulf story is not altogether different from that in Europe. When U.S. and allied interests have intersected, as was the case with Operations *Earnest Will* and *Desert Shield/Desert Storm*, support has been forthcoming. The close ties that the United States had forged with Saudi Arabia and with several of the other Gulf states certainly helped lay the foundation for cooperation—but simple convergence of interests is probably itself a sufficient explanation, with concern over Iraq's behavior and future potential helping elucidate even Syria's willingness to support coalition efforts in *Desert Storm*.

---

[15]Cushman (1987); Tyler (1988a and 1988b); and Cushman (1988).

[16]Woodward (1991).

[17]Wright and Montalbano (1996); Sisk (1996); Lichfield (1996); Bruce (1996); "Saudis Not to Let US Launch" (1998); "The Access Issue" (1998); and Jehl (1998). See also other contemporary press and media coverage.

[18]Jehl (1999).

Because the Gulf War ended with Saddam Hussein still in power, there was some support in the region for permitting some U.S. forces to stay, particularly since that presence has had the sanction of the United Nations. Insofar as actual combat operations against Iraq have been concerned, however, the Saudis and some of their neighbors did not and do not feel comfortable serving as bases for what some see as continued harassment of Iraq. While military action to defend their territories and economies—as had been taken in 1987 and 1991—was acceptable, these more recent strikes have not been seen as advantageous to the host states but have instead been viewed as a potential irritant to Baghdad in a region where grudges can be long-lasting. "You Americans will eventually go home," the Gulf countries in essence say, "leaving Saddam's regime intact and us, his neighbors, vulnerable to retribution." It should not be completely surprising, then, that the Saudis and their neighbors have concluded that they have little to gain from supporting these ongoing and inconclusive U.S. attacks.[19]

The United States has other friendships in the region, but each has posed its own complications. Jordan, a friend of many years' standing and a state that the United States has characterized as a "major non-NATO ally," failed to provide any support during the Gulf War and has opposed several U.S. strikes on Iraq since that time. Amman's close ties to Iraq as a balancer against Syria have on these occasions outweighed its desire for closer relations with the United States. Israel, also a close friend and a major non-NATO ally, is problematic as a base of operations for reasons relating to the continued uncertainties of its regional position.[20] States such as Georgia and Azerbaijan, which are actively courting U.S. friendship and which, like Turkey, span the political geography of Europe and the Middle East, may well be willing to offer additional support. However, their situation is complicated by their relationships with Moscow and by Russia's desire to maintain (or regain) its influence

---

[19]Not surprisingly, the most notable exception to this attitude has been Kuwait, which has supported the majority of U.S. actions. Kuwait, of course, continues to feel the greatest threat from Iraq and thus the greatest security dependence on the United States.

[20]See Khalilzad, Shlapak, and Byman (1997) for a full discussion of these issues.

over former Soviet dominions, as well as by internal challenges to their long-term stability.

Thus, while the Gulf experience with access constraints is not dissimilar to that of Europe, the overall Gulf and Middle East environment is considerably more problematic. Specifically, the lack of strong alliance ties creates a great deal more uncertainty, and failure to ensure the right to use forces located in certain countries, such as Saudi Arabia, continues to plague operations and planning. Moreover, there would appear to be no easy way to ensure that the situation improves in the future. The only obvious solution is to develop alternatives to heavy reliance on any single state such as Saudi Arabia, but this is far from easy in a region whose volatile politics require that U.S. policymakers remain abreast of the nuances of each state's strategic position prior to asking any favors—or assuming the existence of common goals. Here, more effort at convincing friends and allies of the U.S. position may well be in order. Greater transparency regarding U.S. objectives and more extensive sharing of intelligence could help bring others' strategic assessments more closely in line with those of the United States, but such steps need to "begin at the beginning" with full awareness of the limitations imposed by regional concerns of too-strong ties to the Americans. In the meantime, the forces in the Gulf remain in place despite the difficulties encountered in actually using them over the past decade.

Asia[21]

Although the United States maintains a strong and sizable presence in East Asia and the Pacific, it has not been involved in any substantial military operations in that region in some time. The level of U.S. political commitment here, however, is quite high. Bilateral defense agreements with Korea, Japan, Australia, the Philippines, and Thailand and unilateral commitments to provide for the security of the Marshall Islands, Palau, and the Federated States of Micronesia all promise sustained U.S. involvement. Furthermore, the United States maintains a small number of forces in Singapore and has a

---

[21] Khalilzad et al. (2001), Chapter Four, contains a first-order discussion of USAF basing requirements and options.

substantial investment in the development and growth of defense ties with a number of other regional states.

Certainly, there is a fair amount of regional agreement that the United States should remain involved in this part of the world. Although it has been some time since anything other than exercises and the occasional humanitarian operation have actively involved U.S. forces here, many in the region believe that the U.S. presence acts as a stabilizing force. Yet opinions regarding what exactly the United States is expected to stabilize vary. Some Asians fear the emergence of a hegemonic China, while others worry about a rearmed and imperialistic Japan. Japan and Korea, meanwhile, remain separated by centuries of mutual distrust. U.S. security guarantees are seen as hedges against all of these dangers.

At the same time, however, most countries in the region wish to avoid inflaming tensions in what is seen as a fairly stable and highly prosperous period of history. Thus, for example, few are willing to openly avow support to the United States if it comes to the aid of Taiwan in a possible war with the People's Republic of China (PRC). These divergent attitudes and desires make it difficult to predict how countries would respond to U.S. calls for support, as such response is likely to be highly variable and sensitive to the details of the specific scenario.

U.S. forces in Asia are permanently based primarily in South Korea and Japan, with smaller components on the sovereign U.S. territory of Guam and in Australia and Singapore. The latter's strategic interest in maintaining good ties with the United States is self-evident; it has offered increased access for U.S. forces, an offer that is being taken advantage of.[22] While the United States enjoys no permanent presence in Thailand, the two countries participate in a regular and substantial program of military exercises and maintain close ties. Finally, U.S. arrangements with Palau, the Federated States of Micronesia, and the Marshall Islands are based on compacts of free association that commit the United States to take full responsibility

[22]Hua (1998).

for the security and defense of these countries. In exchange, the United States enjoys considerable access rights.[23]

Of course, the United States also has a long-standing program of close cooperation with Australia. As Canberra reevaluates its security situation in light of recent events in Indonesia and elsewhere in Southeast Asia, opportunities may arise to develop this relationship even further.

But it is Japan and Korea that form the cornerstones of the U.S. presence in Asia. U.S. bases in these two countries are substantial, long-standing, and governed by agreements similar to those with NATO allies in Europe. More important, these host countries view U.S. presence as a vital component of their own national security and defense. In the case of Korea, U.S. forces located on Korean soil are there specifically to help the host country defend against attack. As for Japan, Tokyo has proven its commitment and friendship as Washington's key partner in Asia both throughout the Cold War and since its end, and continued limits on the role of Japan's own military strength ensure that the relationship remains mutually advantageous. While there have been domestic concerns in Japan about the scope and impact of the U.S. force presence in their country, revised guidelines for the U.S.-Japan defense partnership, approved in 1997, have further strengthened bilateral military ties. Among other things, these guidelines formalize a commitment on Japan's part to support U.S. forces in the area, as required, for example, during a regional crisis. This builds on prior commitments to bilateral cooperation in support of Japan's own defense. Thus, history, force structure, and formal arrangements all increase the likelihood that both Japan and Korea will continue to view support of U.S. policy as in their own best interests.

That said, as was demonstrated in the discussion of Europe, even highly reliable partners sometimes change their minds. The classic

---

[23]Palau was the last of the three countries to sign such a compact. Although Palauans voted in favor of such a compact in seven successive referenda beginning in 1983, their state's constitutional bar on nuclear materials on its territory was incompatible with the requirements of U.S. access and presence, which might have necessitated transit of such materials through Palauan waters and airspace. A constitutional amendment finally removed this block, and the compact was signed in 1993.

Asian example is the Philippines. U.S. bases in the Philippines were closed at that country's request in 1991 after protracted negotiations focusing primarily on remuneration resulted in an agreement that the Philippine Senate would not accept.[24] Following base closure, relations deteriorated and defense ties between the two countries were largely curtailed by 1996, although the defense commitment remained binding.[25] Now, relations are on the upswing, as evidenced by a newly negotiated status-of-forces agreement and a planned program of exercises. It seems likely that the Philippines' renewed interest in U.S. friendship stems primarily from a desire for U.S. support in its continuing diplomatic—and intermittently military—dispute with China over ownership of the Spratly Islands. The United States, for its part, insists that it takes no position on the Spratly issue, seeking only to protect the sea-lanes of communication (SLOCs) in the region. While port visits and training exercises with the Philippines are resuming, the United States has said that it has no intention of reestablishing a permanent presence there.[26]

Aside from these political concerns, some geographical constraints to U.S. capabilities in Asia also exist. Guam, for example, represents a valuable chunk of sovereign U.S. territory in East Asia, but the island is distant from most likely conflict locations. Similarly, U.S. forces in Korea and Japan, while well situated for their primary mission of deterring North Korean adventurism, are based far away from the Taiwan Strait and South China Sea. Physical access both in those critical areas and farther south is currently limited. Similarly, while Diego Garcia may provide purchase in some instances, its distance from much of the Asian region makes it another necessary but insufficient component for the wide scope of potential future operations.

Given that existing basing arrangements provide only incomplete coverage of the region as a whole, and given the uncertainty of the political dynamics of the region, the presence and security ties that the United States enjoys in East Asia should not be construed as

---

[24]A volcanic eruption that did grave damage to Clark Air Base while negotiations were under way did little to help the Filipino case. See Suarez (1988); Briscoe (1988); Sciolino (1988); Blaustein (1991); Albor (1991); "Philippine Senate Rejects U.S. Base Deal" (1991); and "Manila Says Subic Naval Base Will be Closed" (1991).

[25]Storey (1999).

[26]Storey (1999); Gedda (1999); and U.S. Department of Defense (1998).

ensuring adequate access across the range of possible contingencies. To the contrary, scenarios where access could be a problem can easily be imagined. Indeed, the ambiguity of relations among states in East Asia and the clear and continuing U.S. interests there make for a dangerous level of uncertainty with regard to future needs and whether it will be possible to meet them. The only viable solution appears to be to diversify and to hedge, maintaining and building as wide a network of ties as possible so as to increase the odds of access and thus facilitate whatever operations may be necessary in the future.

To a large extent, the United States appears to recognize this. Its broad efforts to engage the wide range of Asian states serves a number of policy goals, and access is certainly among them. Even U.S. efforts to build defense ties with China, which have fluctuated in lockstep with overall Sino-U.S. relations but have also yielded some results, have implications for access. Ongoing contacts include reciprocal naval visits, a number of high-level meetings, and Chinese agreement to allow some continuing access to Hong Kong, long a favorite port of call for U.S. servicemen and women and useful for the refueling and servicing of aircraft on long voyages. Military contacts have also been increasingly pursued with Malaysia, and ties with Indonesia have a long history—although the latter were curtailed in the late 1990s due to events in East Timor.

But if access remains somewhat uncertain in the Asia-Pacific region, it presents even more of a concern in South Asia, where existing U.S. relationships are far less developed than they are in the East even as the region grows increasingly volatile. Unlike the Pacific Rim, this region suffers from poor infrastructure and tremendous poverty. The United States has built some contacts here, but they have proven difficult to sustain. Ties with Pakistan were severely curtailed first in the 1970s and again in 1990 because of U.S. concerns over Islamabad's nuclear ambitions. Pakistan's 1998 nuclear weapon testing confirmed these worries and further strained relations with the United States. The military coup in October 1999 further complicates matters. U.S.-Pakistan cooperation in Operation *Enduring Freedom* has improved the overall climate of relations between Islamabad and Washington, but the tense standoff between Pakistan and India over the status of Kashmir remains a troubling element, and the long-term prospects for relations with Pakistan remain uncertain.

India, which also carried out nuclear tests in 1998, has never been an ally of the United States. Although some ties were built in the early Cold War years, they soon deteriorated to near-nonexistence for most of that period. Contacts were beginning to develop in the mid-1990s, with some exchanges of high-level visits having taken place, when they were derailed by India's atomic testing.

While neither India nor Pakistan seems a particularly likely partner in the short term, the rest of the region is even less appealing. Thus, the possibility of improved relations with India and Pakistan should be left open, particularly if these countries make progress toward defusing the tense situation that prevails between them and take steps to stabilize their nuclear competition.

Another alternative may lie somewhat to the north. Through the PfP and bilateral cooperation programs, connections are being built with several of the post-Soviet Central Asian states, notably Kazakhstan and Uzbekistan, both of which participated along with the United States in the 1998 CENTRASBAT (Central Asian Peacekeeping Battalion) exercise, held "in the spirit" of the PfP. Here, as with the European PfP partners, there are concerns about Russian reactions should ties with the United States deepen too quickly.[27] These post-Soviet states, although carrying considerable baggage of their own, could provide infrastructure and may be worth exploring as potential operating locations should need and opportunity intersect in this area.[28]

---

[27]Russia itself, however, may prove a useful partner in Asia. Military contacts between the Russian Far Eastern forces and the U.S. Pacific Command, for example, have developed substantially over the past few years. Although these contacts have recently been scaled back as part of a general Russian moratorium on military ties with the United States and NATO, there is hope that they will yet be revitalized.

[28]Relations with several former Soviet republics in Central Asia deepened dramatically during Operation *Enduring Freedom*.

Latin America

On paper, the United States has a significant security commitment to Central and South America. The 1947 Rio Treaty created a *de jure* defensive alliance in much of the hemisphere, with each signatory committed to seeing an attack against one as an attack against all. In practice, however, the collective security clauses of the treaty are unlikely to be effectively invoked.[29] At the same time, many of the countries on the American continents share a strong interest in regional security.

U.S. involvement in Central and South America in the 1990s has focused overwhelmingly on drug interdiction, an area of mutual concern to Washington and many regional governments. Supporting contingency access for this mission is a network of partnerships and contacts the United States has built, some specifically for this purpose. Close ties exist with Colombia, Ecuador, Honduras, and El Salvador. Also involved in programs of military exercises and training with the United States are Guatemala, Chile, Bolivia, and the Dominican Republic, and there are plans to initiate some training for Nicaraguan officers in the future. The Bahamas also provides support for the antidrug mission, and Jamaica hosts U.S. and Canadian training exercises. Guyana has declared that the United States is welcome to its airspace and waterways in connection with the drug war, and an agreement to that effect also exists with Trinidad and Tobago. While Venezuelan cooperation has been more variable, it has included some training for host-nation forces. Going far beyond the drug enforcement mission, Argentina—which alone among Central and South American states sent forces to the Gulf War—has been accorded the status of major non-NATO ally.

The United States sponsors multilateral exercises both in the Caribbean and in South America. For example, Belize, Guatemala, Honduras, Nicaragua, El Salvador, and the United States all participated in the "Allied Forces 97" peacekeeping exercises. In addition to the drug war, the United States has undertaken humanitarian ef-

[29]The Rio Treaty has been invoked by the United States (in support of U.S. involvement on the side of El Salvador in its war with Nicaragua and in opposition to the Russian deployment of nuclear missiles in Cuba in 1962), but its actual strength is highly questionable.

forts in the region; in 1998, U.S. personnel were dispatched to Honduras, Guatemala, and Nicaragua to provide disaster assistance following the decimation of the area by Hurricane Mitch.

While facilities and infrastructure in this region are acceptable, other difficulties have emerged to hamper access. The most damaging to the counterdrug mission has been the expiration of the U.S. agreement with Panama to maintain its long-standing forward presence there. The counternarcotics mission effort had relied heavily on the Panama bases; their closure in 1999 greatly limited U.S. ability to monitor the area, cutting coverage by about two-thirds.[30] While some of the resulting slack has been picked up by units operating from Key West and Puerto Rico as well as by the establishment of new facilities in Ecuador, Aruba, and Curaçao, the loss is significant, and the arrangements now in place are far from permanent.[31] Additional difficulties emerged when it became questionable whether Venezuela would continue to grant the United States overflight rights, as loss of these rights would significantly limit the utility of the Ecuador base.[32] The refusal of states such as Venezuela to cooperate on some fronts is indicative of the general ambivalence many in the region feel toward the United States. Costa Rica flatly refused access to its territory in support of counternarcotics operations, for example,[33] and Brazil has avoided the sorts of cooperation agreements that the United States has signed with Colombia, Venezuela, Peru, and Bolivia.[34] While not openly hostile to the United States, Brazil is concerned that the drug mission could end up a cover for U.S. "imperialism" and fears that U.S. agreements with its neighbors could inadvertently push drug traffickers into its territory.[35] Furthermore, government support of U.S. actions may or may not translate into favorable public opinion; Colombians are of mixed mind about U.S. involvement in their country.[36]

---

[30]Abel (1999); Farah (1999).

[31]Grossman (1999).

[32]Abel (1999).

[33]Chacon (1999).

[34]"Armed Forces to Join Drug Enforcement Effort" (1996).

[35] Ibid.; Heyman (1999).

[36]Ibid.

Although U.S. involvement in Central and South America is now heavily focused on drug enforcement, this has not always been the case, as evidenced by U.S. involvement in civil wars in El Salvador and Nicaragua. But both historical conflict and continuing distrust reflect the general uncertainty of this region, the mixed feelings many Latin Americans have about their massive "Yanqui" neighbor, and the difficulty of making clear predictions of the forms future U.S. involvement might take.

In terms of access, the United States has little to be concerned about as long as its efforts are focused north of the Equator. Bases in the southern United States and especially in Puerto Rico provide good coverage of Central America and the northern half of South America. Having lost the bases in Panama, U.S. presence further south is somewhat sporadic. In addition to the newly negotiated arrangements with Ecuador, Aruba, and Curaçao, the United States maintains facilities in Honduras (which were initially established in support of U.S. involvement in El Salvador in the 1980s), and Peru has agreed to host a radar surveillance site. But this presence is minimal compared with the forces that the United States long maintained in Panama. Whether countries that agree to assist the United States in drug interdiction will be as amenable to other undertakings is a question that has not yet been satisfactorily answered.

The danger that U.S. actions will be interpreted as imperialistic requires that particular care be taken in engagement in this region. Transparency of goals and structures is important here, but it is equally important to strengthen ties in peacetime and to continue to build economic relationships that can help foster trust. Military ties alone will likely not be sufficient to alleviate local concerns and may backfire in the long run, given the history of human-rights abuses associated with cooperation between U.S. and Latin American militaries.

Africa

Like South Asia, Africa is something of a void for U.S. engagement. During the Cold War, close military ties existed with Somalia and Kenya, mostly to facilitate access to Southwest Asia. Today, Somalia is in ruins and relations with Kenya have cooled, although they remain nominally friendly.

The lack of U.S. involvement in Africa has also created something of a vacuum in understanding the complex political realities that drive relations between states there.[37] Furthermore, Europe, the United States' most reliable security partner, is linked to Africa by a long and painful colonial and postcolonial history that further complicates the equation.

What the United States or anyone else can do in Africa is significantly constrained by the abysmal infrastructure and dearth of sophisticated local forces to contribute to operations, humanitarian or otherwise, on the continent. While South Africa maintains a highly modern and effective military, it is located at the southern tip of the continent and has shown little inclination to cooperate with the United States in ventures such as the building of a U.S.-sponsored African peacekeeping force.[38]

The current outlook for Africa suggests that U.S. operations there will focus on peace enforcement, response to humanitarian crises, or both. However, large-scale operations of these kinds could be difficult to execute given the region's woeful infrastructure and the long distances between where U.S. forces would come from and where they would need to go.

Continuing to support regional peacekeeping may be a means of limiting the need for substantial direct involvement in African conflicts; however, countries such as South Africa must be persuaded to take part if these endeavors are to bear much fruit. Efforts to engage Johannesburg are an important step forward in this regard.[39] Such multinationalization of peacekeeping would also help temper the dangers of ethnic bias that native African peacekeeping efforts such as ECOMOG (West African Peacekeeping Force) have encountered in the past. Ensuring that such regional forces also have a capability to respond to humanitarian emergencies would be beneficial as well, although the need to move food and supplies large distances may still require the participation of more advanced Western forces.

---

[37]Some of those complexities are discussed in Chapter Four.

[38]Heyman (1999).

[39]Kozaryn (1999).

IMPLICATIONS

As can be seen, even this brief survey of the global access question has managed to raise a bewildering array of issues. And because of the fundamental reality of national sovereignty, many questions planners would dearly love to see resolved—such as "Will Japan give the United States access if China attacks Taiwan?" or "Can we rely on Saudi Arabia to permit USAF operations if Iraq fails to comply with this or that UN resolution?"—elicit responses that can at best be described as hedged. Nonetheless, we believe that our analysis of the past and present record of U.S. overseas access does allow for the elucidation of some general principles. In concluding this chapter, we therefore wish to put forth a set of six factors, three of which seem to increase a partner's cooperation with the United States and three of which work against such cooperation. The three "pros" are

- Close alignment and sustained military connections;
- Shared interests and objectives; and
- Hopes for closer ties with the United States.

The "cons" are

- Fear of reprisals;
- Conflicting goals and interests; and
- Domestic public opinion.

We will briefly discuss each of these six in turn.

Close Alignment

It should come as no surprise that states with long-standing security relationships with the United States will, all other things held equal, be more likely to support U.S. actions. Probably the best example is Great Britain; the "special relationship" that London and Washington have cultivated over the past 60 years has paid great dividends for the United States. Alone among U.S. allies, for example, Britain supported the U.S. strike on Libya, and British forces flew alongside U.S. aviators in Operation *Desert Fox*.

At the same time, it must be noted that an alliance relationship is by no means a panacea. We have already noted that on many occasions NATO members have denied access to the United States, sometimes with serious consequences.[40] Nonetheless, the United States' worldwide web of security arrangements—alliances, treaties, and understandings—has been and will continue to be an integral part of any global access strategy.

Shared Interests and Objectives

Again, shared interests and objectives obviously favor cooperation with the United States. Even friends as notoriously prickly as the Saudis, for example, have extended a warm welcome to the United States when their understanding of both the situation at hand and the steps needed to deal with it has coincided with that of the United States. It should be noted, however, that agreement needs to cover both means and ends. Riyadh, for example, may want to see Saddam's regime deposed even more than does the United States; however, if the Saudis see Washington's desired strategy as ineffectual or counterproductive, they are unlikely to cooperate even in pursuit of a shared goal.

Furthermore, confluence of interests in a specific situation should not be seen as translating into congruent views in other instances. If nothing else, the preceding analysis should demonstrate that each government considers the granting of access on an immediate, case-by-case basis. Certainly access is more likely to be granted when interests coincide, but as a situation evolves, views may evolve as well, and perspectives once shared may thus be shared no longer.

Greater transparency and information sharing can be powerful tools of persuasion for the United States, just as they were when intelligence regarding Iraqi troop movements helped convince the Saudis to accept U.S. forces after Iraq's 1990 invasion of Kuwait.[41] Transparency and information sharing in general, even when no crisis is looming, can help ensure that states worldwide have a better

---

[40]The events surrounding Operation *Nickel Grass* in 1973 indicate that even the UK will not automatically support U.S. actions.

[41]Woodward (1991).

understanding of U.S. goals and motivations. This can help remove suspicions of hidden American agendas and convince others that their interests are in harmony with those of the United States.

Hopes for Closer Ties with the United States

Our analysis suggests that the old adage about "friends in need" holds true in contemporary international politics. Countries looking to improve their relationships with the United States[42] or perceiving their security to be closely dependent on U.S. support[43] may be particularly prone to providing access.

While close friends like the UK may be inclined to support U.S. initiatives, mutual treaty commitments do not ensure such cooperation. In fact, actors like Greece may represent the opposite side of the coin. Confident that their actions will not compromise their position in an existing alliance that they know the United States prizes, they may have little incentive to respond affirmatively when the United States asks for assistance outside the narrow bounds of existing treaty commitments. Indeed, as was discussed earlier, Greece cooperated in only a limited fashion during Operation *Allied Force*.[44]

At the same time, countries hopeful of improved relations with the United States appear somewhat likely to believe that their support of U.S. efforts now will help ensure U.S. military assistance later. Whether programs such as the PfP that promote ties with the Romanias and Philippines of the world actually translate into eventual U.S. assistance is an open question. In the meantime, these states may be likely to grant access and support for a range of operations.[45]

---

[42]Portugal in 1973, Hungary in 1999.

[43]Kuwait since 1991, perhaps the Philippines today.

[44]It is likely that Greece's response also had something to do with its interminable confrontation with Turkey over Cyprus. To the extent that the Greeks cannot count on U.S. support in resolving this immediate bone of contention, they had still less motivation to support U.S. actions outside the strict legal limits of NATO's charter.

[45]India's enthusiastic support for U.S. operations in Afghanistan in 2001–2002 almost certainly owes something to Delhi's desire for improved ties with the United States, as

Fear of Reprisals

Among the factors that can work against other countries granting access to the United States is a fear of reprisals. Britain, Spain, and other European actors refused to provide access and overflight for the 1973 airlift to Israel because of concerns over economic retribution from Arab states. And Portugal, the one country that did support the U.S. airlift, was indeed subjected to a cutoff of oil from the Persian Gulf. In 1986, many of the same countries as well as France were concerned that a barrage of terrorism might be directed at them if they cooperated in *El Dorado Canyon*. Today, fears of possible reprisals certainly figure in many Gulf Arab states' reluctance to support U.S. raids on Iraq.

In many cases there may be little or nothing that can be done to assuage these concerns, as the United States has had little enough success battling terrorism itself and is seldom in a position to insulate its partners from the effects of economic sanctions. At the same time, the United States can offer to help protect the host country from direct military retaliation such as air and missile strikes or outright invasion. And by sharing intelligence and threat assessments with the host government, Washington may be able to provide some reassurance that the consequences of cooperation will be relatively minor. That said, friendly countries' fear that adversaries might strike back at them will remain a barrier to cooperation.

Conflicting Goals and Interests

Just as shared objectives can facilitate access, so too can interests that are not congruent destroy prospects for cooperation. This factor has played heavily in Saudi behavior since the Gulf War and contributed to Turkey's reluctance to support proposed U.S. action when Iraq launched its offensive against the Kurds in 1996. Greece and Macedonia's refusal to lend full support to NATO's war over Kosovo and Serbia was similarly based at least in part on different images of "stability" in the Balkans. As was suggested earlier, trans-

well as to the opportunity the "global war on terrorism" presents to recast Indian operations against Kashmiri militants in a new and favorable light.

parency and information sharing are the primary tools at the United States' disposal in combating this problem.

Domestic Public Opinion

Even governments that are not true democracies are usually sensitive to the tides of public opinion; it is, after all, better to be a popular dictator than an unpopular one.[46] And since most U.S. security partners have governments that are at least somewhat answerable to their populaces, grassroots opposition to cooperation with the United States can suffice to stymie even the best intentions of a friendly regime. It was Spanish popular opinion that resulted in the eviction of the 401st TFW from Torrejon, and it is the Okinawan people who have persistently agitated for the reduction or termination of U.S. presence on that Japanese island. And Saudis are sensitive to Islamist complaints that ongoing U.S. presence is inconsistent with Riyadh's role as guardian of Mecca and Medina. Here again, maintaining clear lines of communication and upholding a reputation for honesty and plain dealing probably represent the best weapon the United States has against this impediment.

In sum, then, our survey suggests that there are two fundamental tools available to the United States that are particularly appropriate to help ensure access. The first—*transparency and information sharing*—can help convince friends and allies that their interests do not in fact conflict and that cooperation with the United States aligns with their own goals. The second, *engagement*—which is directed mainly at states where ties are less clear and less strong—helps establish the United States as a good friend to have in one's corner and thus someone for whom doing an occasional favor may be wise. Maintaining an active program of military-to-military contacts and using U.S. "information dominance" to help shape the perceptions of partner countries and other aspects of engagement may be the best assurance that U.S. military forces can find adequate access to perform their missions both quickly and safely when need arises.

---

[46]The Shah of Iran, Anastasio Samoza, and "Baby Doc" Duvalier are just three of the former leaders who would attest to the truth of this.

That said, future access can never be guaranteed—for countries will in the end base their decisions largely on the constraints of the moment. Thus, while the United States can influence such views and make them more amenable to the granting of access—and, indeed, should seek to do so whenever possible—it must be prepared for the failure of even the closest relationships to provide the access it seeks for a given operation. As a result, exclusive reliance on friendships and extant relationships is an error. Rather, the policies of transparency and engagement should be accompanied by increased flexibility of operational and deployment options in order to broaden the choices available to the United States.

This analysis has shown that access is likely to be most troublesome in two regions that are critical to U.S. national security: the Persian Gulf and Asia outside the immediate vicinity of the Korean peninsula. In addition, sub-Saharan Africa and Latin America—particularly in the far south—will pose serious operational challenges. In these areas and probably elsewhere as well, situations will almost certainly arise in which USAF forces will confront missions that must be undertaken with less-than-optimal access and basing. In the next chapter, we will discuss the operational constraints such circumstances can impose and will propose some ways of ameliorating them.

Chapter Three

OPERATIONAL CONSIDERATIONS AFFECTING ACCESS REQUIREMENTS

OVERVIEW

We set out to evaluate how less-than-optimal access—by which we mean, in essence, basing farther away from the target area than is standard USAF practice—would affect an expeditionary force's operational capabilities.[1] Toward this goal, we explored air expeditionary task force (AETF)[2] operations in a notional major theater war: an Iranian attack on Kuwait.[3] For this analysis, we

[1]Our interest in this problem should not be interpreted as recommending that distant basing should be the default or preferred model for USAF operations. Nor are we suggesting that accepting a suboptimal beddown is the only option available to a commander should some combination of threat, political restrictions, and/or infrastructure limitations create difficult basing choices. Air Force planning and operations need to be sufficiently flexible and robust to permit rapid, effective operations in challenging circumstances, including the possibility that the fight may need to be undertaken, at least initially and for some period of time, from distant bases.

The authors did not have access to detailed information regarding extended-range fighter operations during Operations *Allied Force* or *Enduring Freedom*. However, discussions with knowledgeable people within the USAF suggest that the "real world" experience has proven broadly consistent with our analysis. Indeed, if anything, our work may be somewhat optimistic regarding the operational consequences of remote basing for shorter-range assets.

[2]AETF is the USAF name for a forward-deployed force package. An AETF can draw on assets from "in-the-box" AEFs as well as from other Air Force units as needed.

[3]This scenario is simply a vehicle for exploring a set of requirements and thus involves only (1) a notional campaign that demanded a full set of USAF operational capabilities in response, and (2) a threat that rendered sustained operations from forward bases a potentially risky proposition.

- Identified potential basing options for both the fighter and support elements of the deployed force;
- Selected alternative pairs of beddown locations (one base for fighters and another for support assets) to study the impact of increased distance between bases and targets;
- Employed a sortie-generation model to estimate the force's combat capability from each set of bases; and
- Adjusted key parameters determining operational effectiveness and repeated the process.

Our work focuses on the effects of being forced to base at more distant locations because of enemy offensive capabilities that seriously threaten closer-in facilities. However, the effects we identified and the remedies we recommend would be equally applicable to a situation in which political constraints led to limited basing options.

Figure 3.1 shows the methodology we used.[4] Factors shown in shaded circles are explicitly considered in our analysis, while those shown only in outline are exogenous variables about which we made assumptions.

Support and fighter deployment bases were selected on the basis of a comparison of available installations in the area of interest and on the basis of critical aircraft operating requirements such as minimum runway length, munitions storage and handling facilities, parking ramp space, fuel storage capacity, and so on. Having postulated threat capabilities—we assumed the adversary had numerous 550-nm-range surface-to-surface missiles—we chose two sets of airfields for the deployed forces:[5] one close in and at risk of enemy attack and the other outside the assumed range rings of the opponent's missiles.

---

[4]Antimissile defenses were not factored into this analysis. If the United States possessed robust, deployable, and highly effective antimissile capabilities, one constraint on basing options would be largely removed. However, even if defenses effectively reduced threat capabilities to zero, geography, air-base characteristics, aircraft operating requirements, and the political factors discussed elsewhere in this report could still interact to compel USAF expeditionary forces to operate from extended range.

[5]The Chinese M-18 was chosen as the nominal ballistic missile threat. Note that this is not a "worst case"; the North Korean *Nodong*, for instance, is assessed as having a range of some 700 nm.

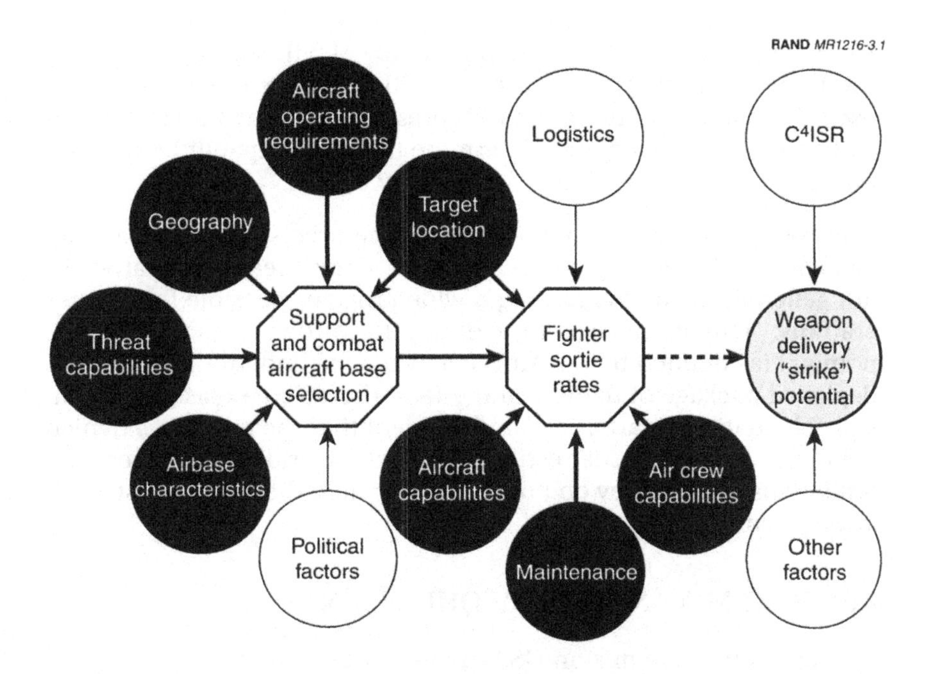

Figure 3.1—Analytic Methodology

The support and fighter bases selected were combined with target location, aircraft capabilities (speed, range, etc.), maintenance capability, and aircrew capabilities (we used a maximum sustained fighter crew duty day of 12 hours) to model fighter sortie rates. These sortie rates were then used to estimate the combat power and support requirements (such as fuel and munitions required per day) for a deployed force package. Availability of adequate logistics; command, control, communications, computers, intelligence, surveillance, and reconnaissance (C^4ISR); and factors such as enemy action, weather, and the like were not considered in this analysis.

Limitations of This Analysis

This work does not purport to incorporate all of the myriad factors that are involved in planning or executing real-world combat operations nor to represent the full gamut of issues that affect USAF expeditionary operations. Every scenario is unique, and the number and types of forces deployed and employed will vary substantially from one to another, as will the specifics of how those forces are used. We have not attempted to represent the many important operational details that would determine the exact capabilities and requirements of any individual contingency.

However, being "exactly wrong" about the precise operational and tactical specifics of any one scenario does not mean the analysis is not generally correct regarding a wide range of possible future operations. Our intent was to estimate the broad capabilities of a nominal (as defined by the Air Force when we undertook the study) deployed package of fighters and attack aircraft in a reasonable but still-schematic scenario. We are confident that the results we depict are reasonable first-order estimates of "real-world" capabilities and limitations, even if they do not precisely match the characteristics of any particular case.

AIRCRAFT MIX AND BASE REQUIREMENTS

We based our force mix on USAF plans for a "nominal" AETF of approximately 175 aircraft, as shown in the third column of Table 3.1. We calculated basing requirements and combat capability for the force shown in the fourth column, which represents a typical initial deployment package according to the Air Force.[6] Although our results are based only on this 48-fighter package, the methodology can be applied to any arbitrary force size.

USAF heavy bombers—B-52s, B-1s, and B-2s—are likely to play a prominent role in any power-projection scenario. Their long range makes it possible for them to operate from great distances; bombers based in the United States have regularly participated in strikes on targets in Iraq, Serbia, and Afghanistan. If the situation permits, they can also be forward deployed, which increases the number of sorties they can fly in any given time period.[7] And, as the bomber fleet is

---

[6]Data on the planned AEF/AETF structure comes from Cook (1998). Other so-called low-density/high-demand (LD/HD) assets, such as U-2 and RC-135 reconnaissance aircraft, would almost certainly be tasked to support the AETF. We did not assess the beddown or support requirements for these aircraft.

[7]B-52s have, for example, operated from bases in Saudi Arabia and Oman.

Table 3.1

Notional AETF Composition

Aircraft Type	Role	Total	Initial Deployment
F-15C	Air-to-air	24	18
F-15E	PGM strike[a]	24	10
F-16CJ	SEAD[b]	18	8
A-10	Antiarmor	26	12
E-3	Surveillance	3	3
HH-60	CSAR[c]	12	3
C-130	Airlift	18	8
KC-10	Tanker	6	4
KC-135	Tanker	20	6
C-21	Transport	9	3
B-52/B-1	Bomber	6	0
B-2	Stealth	3	0
F-117	Stealth	6	0
Total		175	75

[a]PGM = precision-guided missile.

[b]SEAD = suppression of enemy air defenses.

[c]CSAR = combat search and rescue.

modernized, it will be better able to deliver precision weapons in all kinds of weather. That said, this analysis focuses on how the USAF shorter-range fighter-bomber force can be affected by access restrictions, and we do not consider bomber basing or operations. To the extent that the tanker resources deployed with the AETF would be used to support bomber operations, however, this analysis will understate the number of refueling aircraft required.[8]

We began the process of determining basing requirements by calculating the minimum runway length requirements for each aircraft type. We started by defining typical operating configurations for each fighter and attack aircraft type: mixes of fuel, munitions, external tanks, and so forth. Configurations included those most commonly flown in current USAF no-fly-zone enforcement sorties over Bosnia and Iraq. In addition, we included options suitable for more

---

[8]In a similar manner, we do not fence off tanker assets to support Airborne Warning and Control System (AWACS), Joint Surveillance Target Attack Radar System (JSTARS), strategic airlift, or LD/HD operations.

intensive air-to-ground operations such as Operation *Allied Force.* We investigated several gross weight and drag configurations for F-15Es and F-16Cs while using a single configuration each for A-10s and F-15Cs. Because we could not predict either the exact missions (strike, close air support) that deployed USAF fighters will fly in the future or the weapons they will be called upon to carry, we conservatively used the highest weight and drag configurations for each aircraft to calculate takeoff and landing data.[9]

Takeoff and landing data (TOLD) for each aircraft configuration were computed using the flight-planning tables from the appropriate USAF technical order ("Dash-One"). In all cases takeoff was calculated for 2000-foot pressure altitude at 32° Celsius with a 1 percent uphill gradient. Using these assumptions and standard USAF criteria,[10] we calculate that a runway of at least 8200 feet is required to accommodate a deploying fighter unit that includes A-10s. The F-15C/E and F-16C can safely fly from slightly shorter (7500-foot) runways.[11]

We also computed how much ramp space would be required to park the 48 fighters in our forward-deployed force package. Again using USAF planning factors, the total comes to about 360,000 square feet; an additional 200,000 or so square feet would be needed to support C-130 operations into and out of the base, making for a total of 560,000 square feet (or 200,000 square feet and 48 available shelters).[12] Finally, the base must also have fuel storage facilities, water, and a munitions storage area.

---

[9]F-15E gross weights ranged from just under 74,000 lb to nearly 79,000 lb with drag indices between about 82 and 97. F-16 gross weights ranged between 35,000 and 40,000 lb with drag indices between 110 and 170. A-10 gross takeoff weight was assumed to be about 43,000 lb with a drag index of about 6.5, while F-15C gross takeoff weight was estimated at about 55,000 lb with a drag index of 51.

[10]From U.S. Air Force Handbook, AFH 32-1084, Chapter 2.

[11]Weather conditions such as heavy rain or ice on the runway could significantly increase landing roll and therefore the required runway length. The calculations above assume a dry runway. Standard USAF procedures prohibit operations when flight conditions will result in the planned use of more than 80 percent of the available runway.

[12]Fighter parking space is based on requirements in AFH 32-1084, Table 2.6. Details of this calculation can be found in the appendix.

The support aircraft assigned to our nominal AETF differ from the fighter and attack jets in a number of ways. The most obvious is that many of them—the KC-10s, KC-135s, and E-3s—are much larger and heavier. In addition, they generally have much lower thrust-to-weight ratios at higher gross weights than do fighters. These characteristics lead to different operating characteristics and correspondingly different base requirements for support aircraft.

Consider minimum runway length. TOLD was computed for fully loaded KC-10s, KC-135s, and E-3 aircraft using the appropriate flight-planning charts and tables and based on the same environmental assumptions used for the fighters. Under these conditions, all three of the heavy support aircraft require very long runways—up to 11,800 feet—to operate at their maximum weight.[13] USAF planning factors call for a minimum runway width of 148 feet for these aircraft. These big, heavy aircraft also require a strong runway; the KC-10 requires a pavement classification number (PCN) of 70, whereas the smaller and lighter KC-135s and E-3s require a PCN of about 50.[14]

Like the fighter aircraft, the support aircraft need some place to park. Since most of these aircraft are too large to tuck into tactical aircraft shelters, they must park on open ramps. Using the same methodology employed for determining fighter parking requirements, we find that the support aircraft in the deployed tranche need some 900,000 square feet of ramp. Adding to this an additional 200,000 square feet to handle airlift loading and unloading brings the apron required to

---

[13]It is possible to operate these aircraft with reduced fuel loads from shorter runways. For example, the USAF operates KC-10 and KC-135 aircraft from Incirlik Air Base in Turkey off runways about 10,000 feet long. However, the reduced takeoff weights required result in less fuel available for transfer to fighters and/or less range and endurance. This is not a critical factor for the Operation *Northern Watch* missions flown out of Incirlik, as the ranges involved are comparatively short. However, as the analysis below will show, as range to target increases, tanker capacity becomes an important constraint on combat sortie-generation capability. Anything that reduces tanker offload capacity—such as operating from shorter runways—directly affects combat power at longer ranges. Therefore, this analysis assumes that commanders will prefer to operate tankers from long runways that maximize their fuel offload capability.

[14]PCN is the standard International Civil Aviation Organization system for reporting pavement strengths; it is determined by an engineering assessment of the runway to determine its load bearing capability. See U.S. Air Force Air Mobility Command (1997), Table 1, for required runway widths and PCNs.

1.1 million square feet. Table 3.2 summarizes the criteria for selecting a viable candidate base for our fighter and support forces.

Table 3.2 shows that on most dimensions the support aircraft basing requirements are more stringent than those for the fighters. The force could therefore deploy to a single large base that met the overall requirements for the support aircraft and had some 1.5 million square feet of ramp space, adequate fuel storage and handling to service all of the aircraft together, and munitions storage facilities.[15]

This option is attractive for two main reasons. First, it minimizes the resources required for force protection against ground threats by consolidating operations at a single location. Second, it may improve overall force coordination and effectiveness by allowing fighter and support crews to interact and plan missions face to face. As the analysis that follows will demonstrate, however, few bases meet the combined basing requirements in many areas of the world. Therefore, in many future deployment situations the USAF will find it necessary, as has often been the case in the past, to bed down the combat and support elements of the deploying force packages

Table 3.2

Required Air Base Characteristics

Characteristic	Fighters	Support
Runway length (ft)	8,200	11,800
Runway width (ft)	150	148
PCN	43	70
Parking ramp space (sq ft)	560,000	1.1 million

SOURCE: U.S. Air Force Air Mobility Command (1997).

NOTE: The ramp space required for fighters can be reduced if adequate shelters exist. The minimum necessary ramp space is 200,000 square feet if 48 shelters are available. Fighters can operate from runways narrower than 150 feet, although few 8,200-foot runways are less than 150 feet wide. Indeed, in the example described in this chapter, this constraint had no impact on the number of fighter-suitable fields available.

[15]Again, the ramp space requirement could be reduced if some or all of the fighters could be parked in shelters.

separately. This separate basing approach, while requiring more resources to protect against ground threats, makes more efficient use of regional basing infrastructure and, by dispersing operations, complicates the enemy's missile targeting problem. The analysis we present next assumes that two bases are being used.

MATCHING SUPPLY TO DEMAND: BEDDING DOWN THE FORCE

Having established the necessary air base characteristics to accommodate both the combat and support elements of our deployed force package, we reviewed existing bases in SWA.

In this vignette, we send the initial package of fighter and support aircraft to the Persian Gulf to deter a possible Iranian attack on friendly Arab states. They will need to be based to enable attacks as deep into Iranian territory as Tehran. The beddown decision will also have to take into account a postulated Iranian arsenal of Global-Positioning System (GPS)–guided tactical ballistic and cruise missiles with cluster munition warheads that can pose a threat to parked aircraft and other fixed targets up to 550 nm from their launch locations.[16]

Basing for the Fighter Force

Reviewing fighter base options, we identified 48 regional military or dual-use airfields with runways longer than 7,500 feet. Of these, 34 met all the other criteria set out for basing the fighter force; they are listed by country in Table 3.3.

These bases allow a wide range of options for bedding down the fighter component. However, the physical characteristics of a potential base are not the only operationally significant criteria for selecting a deployment base. Base locations must also be assessed in

[16]See Stillion and Orletsky (1999), Chapter Two, for a detailed description of the sort of conventional missile threat envisioned here.

Table 3.3

Suitable Fighter Bases in SWA

Country	Base
Bahrain	Shaikh Isa
Cyprus	Akrotiri
	Nicosia
	Patos International
Djibouti	Ambouli
Egypt	Bilbays
	Cairo West
Israel	Nevatim
	Ovda
	Ramat David
Jordan	Prince Hasan
	Shaheed Mwaffaq
Kuwait	Ahmed Al Jaber
	Kuwait International
Oman	Seeb International
Qatar	Doha
	Al Jouf
Saudi Arabia	Dhahran
	King Khalid
	Prince Sultan
	Riyadh
	Tabuk
	Taif
Syria	Damascus
	Tiyas
Turkey	Antalya
	Batman
	Cigli
	Diyarbakir
	Erhac
	Erzurum
	Incirlik
	Mus
UAE	Al Dhafra

relation to the kinds of attacks an enemy might be able to bring to bear—in this case, the 550-nm-range missiles fielded by Iran.[17]

[17]A number of countermeasures are available to a U.S. commander confronting an adversary with the ability to strike hard at potential U.S. bases in the area of operations. For example, long-range bombers based in the United States or at other rela-

It is possible that U.S. commanders could choose to base the fighter component of a USAF expeditionary package inside the range of an adversary's missile force, especially if the aircraft could be protected by shelters at the deployment base. However, there are a number of valuable and vulnerable assets—including personnel—that are not so easily protected: aircraft moving to and from shelters, airlifters delivering supplies, maintenance facilities, tent cities, and so forth. Therefore, we further examined the available bases to see which are outside Iranian missile range. Figure 3.2 shows the result, which trims the number of bases to 15.[18]

In Figure 3.2, three bases in Israel are indicated in gray. Attempting to use these bases in a campaign to defend an Arab state can be generously described as problematic, although the evolution of Middle Eastern politics may alter this situation in the future. For now, however, these bases would clearly be on the bottom of the list of options for hosting USAF forces in a scenario such as this—if indeed they were included on the list at all.[19] This leaves us with 12 potential

---

tively "safe" locations could carry the brunt of the initial burden, perhaps focusing their attacks on enemy offensive capabilities to "defang" the opponent and facilitate a secure deployment of shorter-range assets into the theater. In the example that follows, we focus on the option of basing outside the adversary's threat rings, using that as an example of situations where, for reasons either operational or political, the USAF may need to fly and fight from suboptimal locations. This is not to suggest either that threat is the *only* factor that could drive the USAF to conduct operations from distant bases (recall *El Dorado Canyon*) or that remote basing is the *only* option available to cope with an enemy's threat to closer-in targets.

[18]Antimissile defenses might allow USAF forces to operate safely from bases within enemy ballistic missile range. Previous analyses suggest, however, that even a single ballistic missile with a submunition warhead could effectively attack more than two million square feet of aircraft parking ramp space or tent city area (see Stillion and Orletsky [1999], Chapter Two). This represents an area approximately one-third larger than that required to park the entire deployed force package being considered here. Since even a single such "leaker" could be devastating, *effective* airfield (as opposed to city) antimissile defenses would require a system-level probability of kill of very close to 1.0. This level of effectiveness has never come close to being achieved in the *anti-aircraft* mission and is unlikely to be possible at least for many years with *antimissile* systems. Given the growing threat posed by ballistic (not to mention cruise) missiles to airfield operations and the fact that no antimissile system has yet proven itself in either realistic testing or combat, the USAF must retain the option of operating effectively from bases beyond the reach of the most dangerous threats.

[19]See Khalilzad, Shlapak, and Byman (1997) for a discussion of both the political dynamics and the potential military implications of a general Arab-Israeli peace agreement.

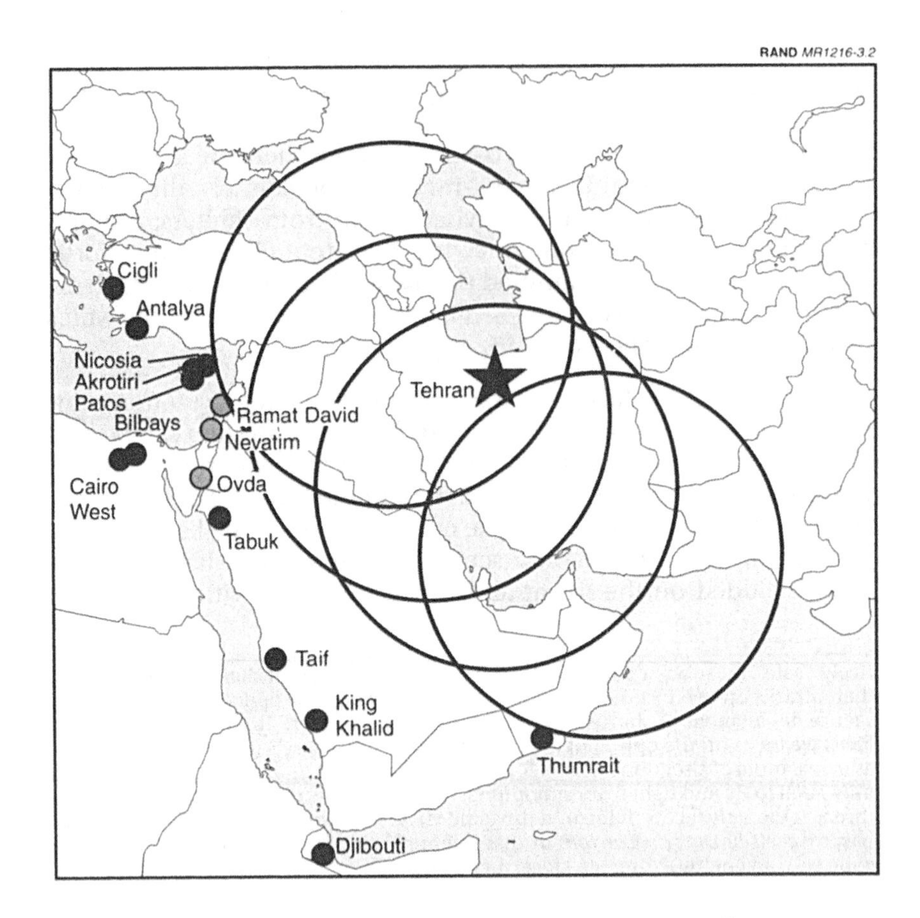

Figure 3.2—Potential Fighter Bases Outside Iranian Missile Range

fighter bases—one-quarter of the original 48. There *are* a large number of potential bases in Southwest Asia; however, once specific operational requirements, threats, and the most elementary political considerations have been applied, the number of realistic deployment options narrows rapidly.

Basing the Support Aircraft

We performed a similar analysis to determine which bases in the region fit the more demanding criteria for support aircraft. Only the 13 installations listed in Table 3.4 qualified; of these, only three—all in western Saudi Arabia—lay outside Iranian missile range, as shown in Figure 3.3. Since AWACS and tanker aircraft cannot be parked in shelters and are critical assets in any air campaign, it would seem essential that they be bedded down at reasonably safe locations. Even more than with the fighters, then, we see that the seeming plenitude of available bases can dwindle in number rather dramatically—from 48 to 3—when attention is paid to the operational environment.

COMBAT CAPABILITY AND BASING

Having determined what basing options existed, we estimated the first-order combat capability of our force package using two different beddown options: one relying on "close-in" bases that could be threatened by the opponent's missiles and another at "safe" locations. This enabled us to identify some critical factors that degrade combat performance at longer range and to evaluate ways of overcoming them.

Table 3.4

Suitable Support Bases in SWA

Country	Base
Bahrain	Shaikh Isa
Oman	Seeb
Qatar	Doha
Saudi Arabia	Dhahran
	King Abdul Aziz
	Prince Mohammed
	Prince Sultan
	Taif
Syria	Damascus
Turkey	Diyarbakir
UAE	Abu Dhabi
	Al Dhafra
	Dubai

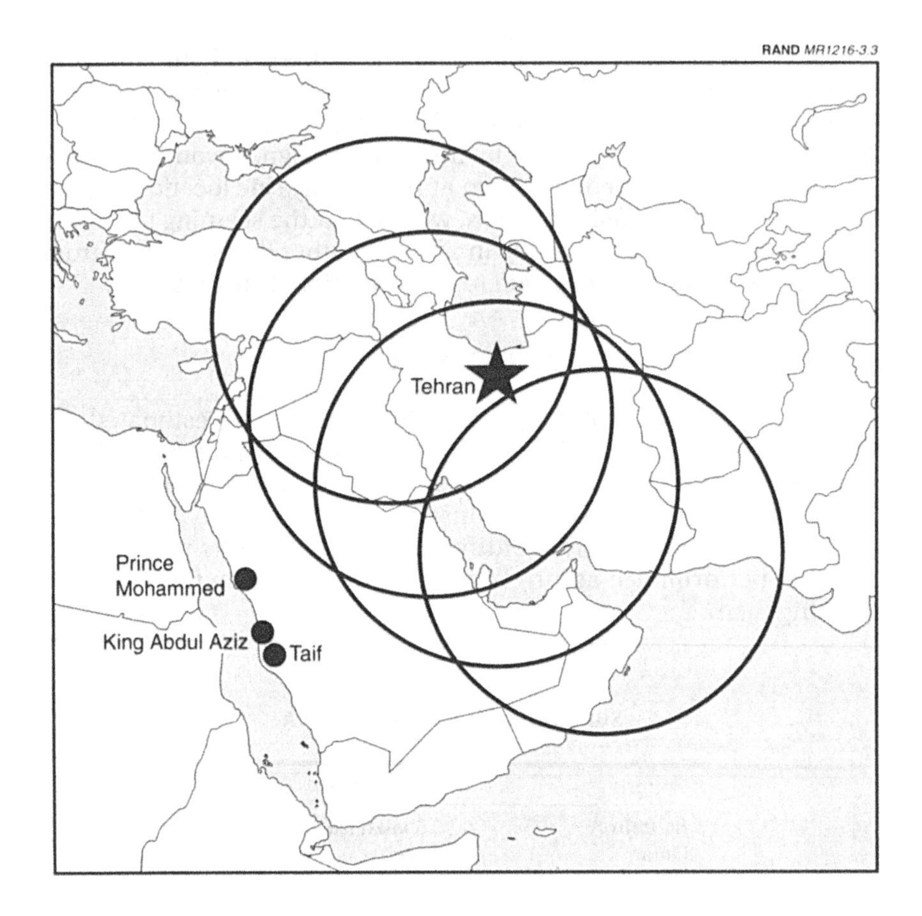

Figure 3.3—Potential Support Bases Outside Iranian Missile Range

Our analysis proceeded in five steps:

- We estimated the unrefueled range for each appropriate aircraft configuration using flight manuals and USAF mission-planning standards.
- We then calculated the amount of fuel that the AETF's assigned tankers could provide each day to support extended-range fighter missions.

- We employed a model to determine the number of sorties the fighters could fly each day from the selected bases to targets in the Tehran area given assumptions regarding aircraft reliability and limitations on both crew and tanker capabilities.
- These results enabled us to calculate in turn the number of strikes that the force could carry out per day from each location.
- Finally, we looked at two major constraints on fighter operations from the more distant locations and evaluated ways of relaxing them.

Estimating Fighter Ranges

We calculated effective combat radii of A-10, F-15C/E, and F-16C aircraft with various combat loads appropriate for missions that involve extended loitering, such as enforcing a no-fly zone or providing on-call close-air support to ground troops. Calculations are based on detailed mission profiles prepared in accordance with individual aircraft flight-planning guidance, tables, and charts as well as with applicable USAF flight-planning regulations. Mission profiles include standard USAF fuel reserves (10 percent or 20 minutes, whichever is greater at 10,000 feet) and enough fuel to fly to a divert base 100 nm from the primary base at 35,000 feet.[20] In general, the A-10 and F-16C configurations we considered require refueling to conduct useful (one-hour minimum time on station) loiter missions beyond about 300 nm from base, while the larger F-15s can generally loiter for an hour or more 500 nm from their base.

We also calculated the effective combat radii of A-10, F-15C/E, and F-16C aircraft with various combat loads for missions involving some low-altitude penetration; the planning materials and factors used for these calculations were the same as those for the preceding ones. Low-altitude profiles are of interest because they are a tactical option that would allow nonstealthy aircraft to attack enemy targets early in

[20]Or 20,000 feet in the case of the A-10. Calculations based on both a fuel reserve and sufficient fuel for a 100-nm divert may appear to be an overly conservative, "belt and suspenders" approach. Because many other factors that could increase actual fuel consumption—such as weather, combat maneuvers, and so forth—are ignored in our analysis, we decided that caution was appropriate.

a conflict while reducing exposure to modern radar-guided surface-to-air missiles (SAMs). However, flying these profiles involves much higher fuel consumption, resulting in shorter combat radii than alternative medium- or high-altitude missions. Employment of long-range standoff weapons could be an alternative to low-altitude tactics if enough of them are available in the early days of a future conflict. Although we calculated a full range of hi-lo-hi profiles for each aircraft type, we found that in general F-15s and F-16s are capable of maximum low-level penetration missions of between 200 and 300 nm. A-10s appeared capable of low-altitude legs of between 180 and 220 nm.

For the analysis that follows, we assumed that fighter aircraft can fly useful missions—whether close air support (CAS)–type missions requiring extended loiter times or hi-lo-hi attack profiles—to a distance of about 300 nm from their base or last air-to-air refueling. We used this average, or typical, distance (except for A-10s) in calculating the daily fuel requirements of a typical deployed fighter force for several reasons. First, while our detailed mission planning for a variety of aircraft configurations (discussed above) shows that different mission profiles and configurations permit somewhat lower or somewhat longer radii, 300 nm appears to be about average for all aircraft types taken together. Second, because it is impossible to predict the precise mix of mission profiles or combat loads a future conflict or peace enforcement operation might require, we opted to use the average, or typical, radii to represent the typical fuel requirements operational commanders could expect under various basing assumptions.

Refueling Capacity and Fuel Requirements

Figure 3.4 presents the offload capability of the six KC-135 and four KC-10 tankers in our nominal forward-deployed AETF elements as a function of the distance they must fly from their base. It assumes that 75 percent of the tankers are mission capable on any given day, is based on flight-planning data in the appropriate technical order, and includes a one-hour fuel reserve.

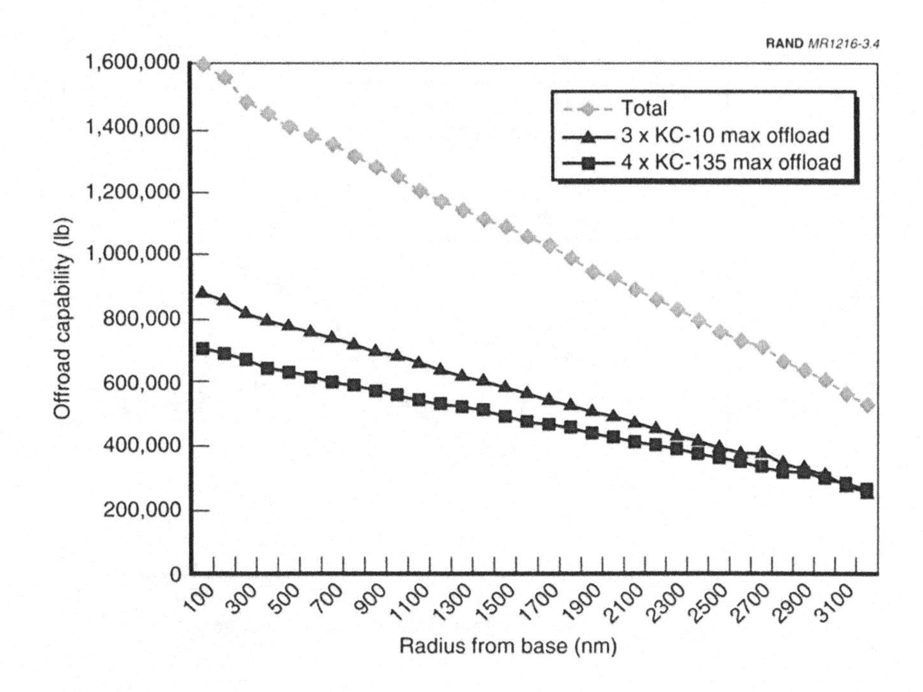

Figure 3.4—Tanker Maximum Offload Capability vs. Range from Tanker Base to Refueling Orbit

In the Southwest Asia example, the tanker force would likely refuel fighters up to about 750 nm from the support base—as close as possible to the Iranian border. At this range, the ten tankers could offload a maximum of about 1.35 million pounds of fuel per day.

Figure 3.5 shows fighter refueling requirement versus the line for total tanker capability from Figure 3.4 for the SWA vignette. Our analysis assumed a 0.8 average mission-capable rate for the fighter force, meaning that 38 fighters flew at any given time.[21] The most distant air-to-air refueling is assumed to take place 300 nm short of the target, and tankers loiter for 75 minutes at this range to begin refueling fighters as they return to base.

---

[21]Fourteen F-15Cs, eight F-15Es, six F-16CJs, and ten A-10s or F-16CGs.

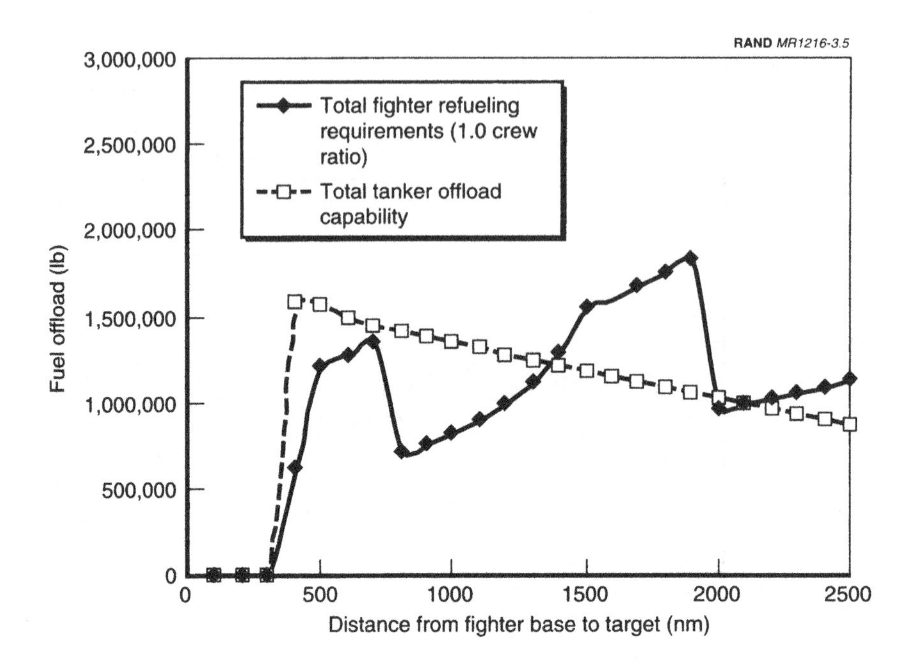

Figure 3.5—Fighter Refueling Requirements vs. Tanker Offload Capability, Southwest Asia

The fighter refueling requirement is zero for missions with radii up to about 300 nm because the fighters can reach these targets unrefueled. Beyond 300 nm, refueling requirements begin to climb rapidly despite the fact that each mission is fairly short because the sortie rate is high. The total refueling required rapidly approaches tanker offload capacity at 700 nm range to target.

For missions longer than about 750 nm to target, crews can no longer fly more than one sortie per day, and refueling requirements therefore drop dramatically.[22] However, they quickly increase once again as increasing mission range drives up fuel consumption while crews are still capable of flying one sortie per day. At the 1300- to 1400-nm point, fighter refueling requirements outstrip the AETF tankers' capabilities. Beyond this point (except for a small region around 2000

[22]As described later in this chapter, crew duty-day restrictions limit sortie generation for longer-range operations.

nm to target where sortie rates drop again), refueling requirements exceed the offload capability of the tankers assigned to the notional AETF depicted in Table 3.1. Either smaller combat forces must therefore be used—resulting in less capability being brought to bear against the enemy—or more tankers must be added.

Augmenting the number of tankers supporting a force is the norm for combat operations. Historical experience, such as the repeated deployment of additional USAF forces assigned to Operation *Allied Force* in 1999, shows that if the assets are available, the USAF will not hesitate to use them even if it means calling up the reserves. Therefore, in small- to medium-size conflicts it is likely that the USAF could project considerable combat power from bases 1500 nm or more from the target area by deploying tanker and crew assets well beyond those included in our nominal force. To minimize disruption across the force, however, prior planning should be sufficiently flexible to accommodate these requirements rather than relying on ad hoc measures devised under the pressures of the moment. Furthermore, in larger conflicts with more capable opponents (i.e., those most likely to have large inventories of accurate ballistic missiles), where the USAF would want to bring the maximum force to bear quickly, tanker requirements (and/or demand for fighter crews, as discussed below) may exceed the available supply.[23]

Estimating Fighter Sortie Rates

Having calculated the unrefueled range of the combat aircraft and the AETF tanker fleet's offload capacity, we can now estimate the ability of the fighter force to produce sorties on a sustained basis. We employed a model that combined the following factors to make this appraisal:

- A regression that predicts the amount of maintenance an aircraft will require after a mission as a function of both cycling components and wear and tear due to continuous use. For example,

[23]Other unanticipated requirements can also place demands on the tanker forces, leading to a relative scarcity of resources. In 2001–2002, for example, numerous tankers were called on to support air defense operations over the U.S. homeland. See Jelinek (2002).

electronic systems tend to break more as a function of being turned on and off (cycles), whereas hydraulic systems tend to break down as a function of how long they have been in use.[24]

- Historic A-10, F-15, and F-16 maintenance hours per flight hour and typical USAF squadron maintenance manning.
- An assumption regarding the maximum allowable aircrew duty day, which we set at 12 hours for sustained fighter operations.
- Available tanker offload capability as described above.[25]

In addition, we have assumed that mission packages are planned to eliminate much of the "orbit" time currently built into many fighter combat mission profiles.[26]

Each of these factors places constraints on the maximum operational tempo (OPTEMPO) USAF fighter units can achieve at a given range to their intended target. Figure 3.6 displays the results[27] and shows that for F-15s and F-16s at very close ranges (less than 200 nm), the ability of maintenance crews to turn aircraft limits sortie production. Between about 300 and 1300 nm from base to target, the limiting factor is crew duty day. Between 1400 and 2500 nm, the limiting factor is the tanker fuel offload capability.

---

[24]See Sherbrooke (1997).

[25]For a complete description of the sortie-rate model used here, see Stillion and Orletsky (1999), Appendix B.

[26]Current mission planning practice calls for fighters to fly to predetermined points and orbit in order to meet up with tankers and to marshal and organize a strike package prior to penetrating enemy airspace. Our mission profiles assume that better coordination of fighter and tanker planning, enroute refueling, and other planning improvements aimed at extending combat radius—perhaps facilitated by a new suite of web-based planning tools—allows fighters to cut total mission orbit time from around 45 minutes to 10 minutes. Since effective ground speed while flying in a circle is zero, this assumption has the effect of allowing crews to reach distant targets more quickly. Current procedures allow an F-15E crew to fly two sorties per day to a maximum radius of about 610 nm. Under our assumptions, the crew could fly two sorties per day to a maximum distance of up to 750 nm. We are indebted to Major Mike Pietrucha and others at HQAF/XOXS for their thoughtful comments and suggestions regarding our fighter mission profile assumptions.

[27]Our model assumes that crews show up three hours prior to their first mission of the day and have a two-hour interval between missions.

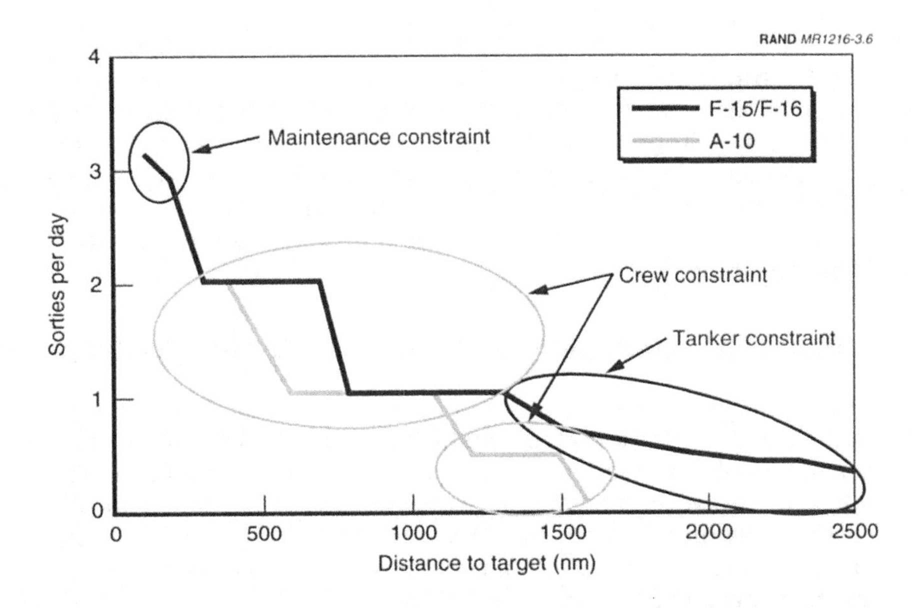

Figure 3.6—Sortie Rates vs. Distance to Target

The story is similar for A-10s. Owing to the much slower cruising speed of these aircraft (about 280 knots vs. some 470 knots for other fighters), however, A-10 sortie rates are generally lower at all ranges because it takes them almost twice as long to complete a mission of a given distance. Crew duty day restrictions begin to limit A-10 sortie rates at shorter distances than they do for the other fighters.

Sortie Rates to Strikes

To illustrate the effect distant basing could have on the force's combat capability, we examined two cases. In the first, we based the package at "close-in" sites: The fighters flew out of Shaikh Isa in Bahrain and the support aircraft were based at Dhahran. For the second, we moved the force outside Iranian missile range, putting the fighters at King Khalid (about 1100 nm from Tehran) and the support assets at Taif.[28] (All four locations are shown in Figure 3.7.) We then translated sorties into strikes for both cases, with the results shown in Figure 3.8.[29] As the figure shows, the number of deliverable fighter strikes drops by over 40 percent, from 88 to 51, when the force is bedded down at the more remote locations.[30] These results stem directly from the three major constraints—maintenance requirements, crew limitations, and insufficient aerial refueling capacity—identified above.[31]

LOOSENING CONSTRAINTS TO RESTORE COMBAT POWER

Of this trio of factors, the limitations imposed by maintenance requirements may be intractable in the near term—at least until a new generation of less maintenance-intensive fighters is introduced. The other two, however, may be more amenable to relaxation. Our discussion now turns to ways the USAF might regain some of the combat capability lost as a result of constraints on crew duty day length and tanker capacity.

---

[28]The fighters could also have been based at Tabuk. We chose to base them at King Khalid even though it is slightly farther from Tehran than Tabuk because King Khalid is much closer to Taif, and the proximity of the two bases would improve coordination and simplify logistics and command and control.

[29]Our definition of a strike is based on sortie, strike, and weapon figures in Cohen et al. (1993), pp. 316, 514, and 531–533. These data indicate that during Operation *Desert Storm* F-117s flew 1299 sorties and conducted 1769 PGM strikes. This works out to 1.36 strikes per two laser-guided bombs. We define a strike as one aircraft delivering approximately 1.36 guided weapons against a target. Aircraft capable of carrying more weapons are assumed to make more strikes per sortie; thus, an F-15E carrying four laser-guided bombs was credited with carrying out approximately 2.9 strikes per sortie.

[30]In the "standoff" case, we replaced the 12 A-10s in the nominal AETF with an equal number of F-16C Block 40 aircraft carrying laser-guided bombs. The A-10's slow cruising speed makes it unsuitable for long-range missions.

[31]These results may exaggerate the impact of longer-range operations on sortie generation potential. After all, not all of the AETF's fighters would be attacking targets in and around Tehran; those flying shorter missions could turn sorties at a higher rate.

Nonetheless, to be most effective as both a deterrent and a "first day" war-fighting tool, the AETF should be able to credibly threaten the full range of targets that an enemy might present. Therefore, we believe that the ability of the force to "go deep" is a valid, if stressful, criterion.

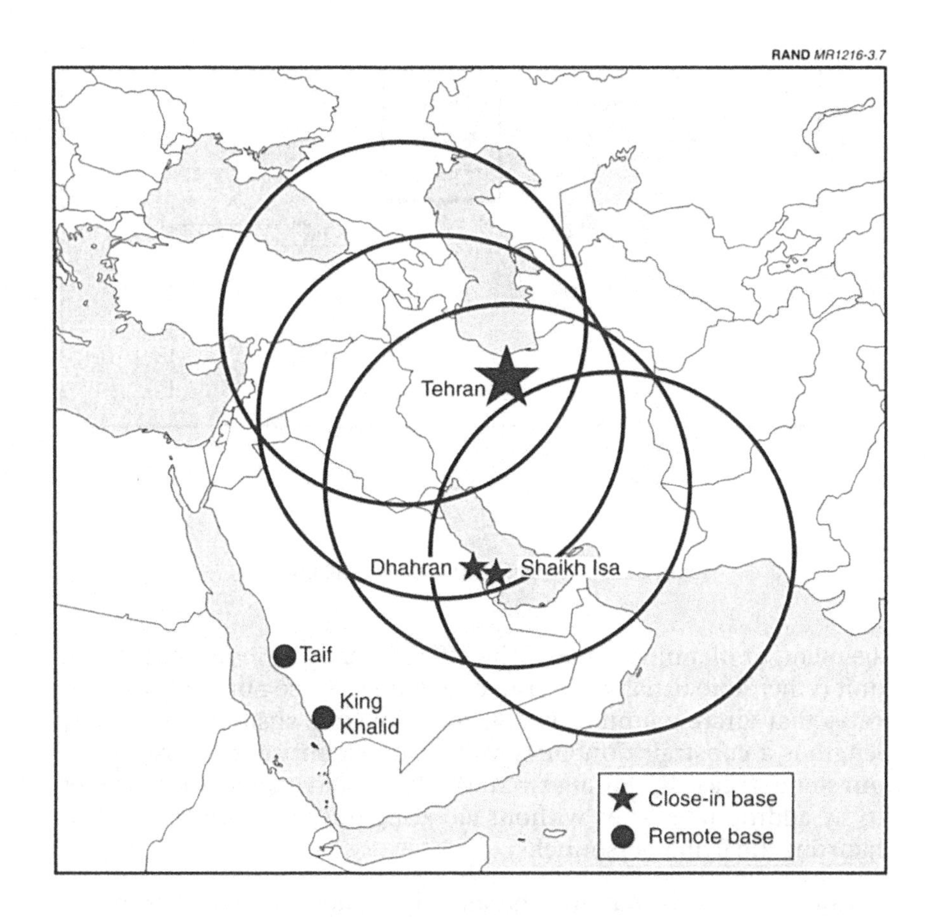

Figure 3.7—Bases Used for Illustrative Analysis

Increasing Crew Ratios *or* Tanker Support

USAF fighter units typically deploy with an overall ratio of about 1.3 aircrew for every aircraft. It would appear from this figure that the standard USAF fighter squadron already has enough crews to considerably reduce the impact of the crew constraints discussed above. However, on any given day a considerable number of fully qualified crews (including squadron commanders, operations officers, operations supervisors, schedulers, and crews assigned to the squadron mission planning cell) will be engaged in essential command, man-

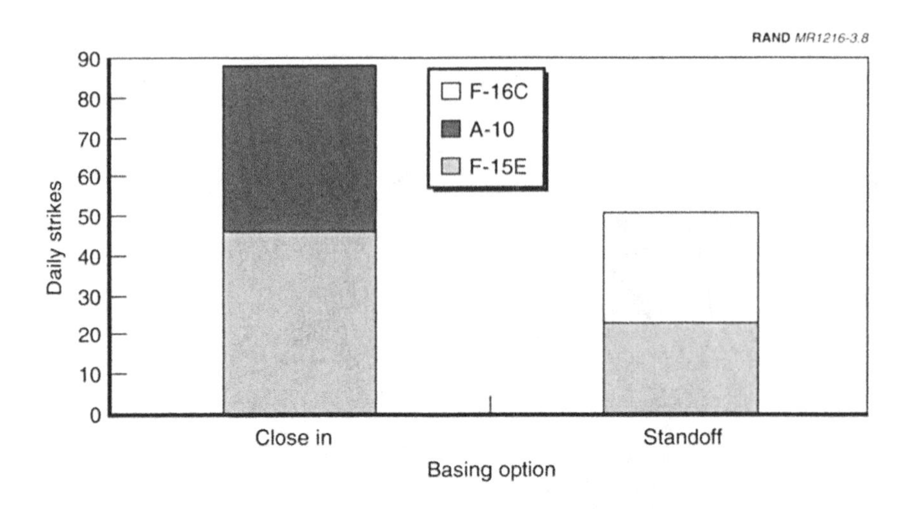

Figure 3.8—Daily Strikes vs. Basing Option

agement, or planning duties. The effective crew ratio for a deployed unit is therefore usually closer to one to one. Since our analysis indicates that aircrew limitations on the allowable sustained duty-day length is a constraint on OPTEMPO, can we improve the results of our standoff case by adding aircrew? This would enable the airplanes to fly additional sorties without violating duty-day norms and exhausting the flying personnel.

Figure 3.9 suggests that, in general, only small improvements in sortie generation can be gained even if the *effective* crew ratio is doubled to 2.0.[32] This is because the deployed force is operating close to its maximum tanker offload capability at most ranges; increasing the number of aircrew by and large means that instead of bumping against a human constraint we instead hit upon one imposed by tanker fuel offload limitations.

Similarly, increasing the number of tankers available to support the AETF does little to improve sortie generation absent an augmented

---

[32]Because of the need for extra rated personnel to perform nonflying functions noted earlier, the actual deployed crew ratio would be higher than 2.0.

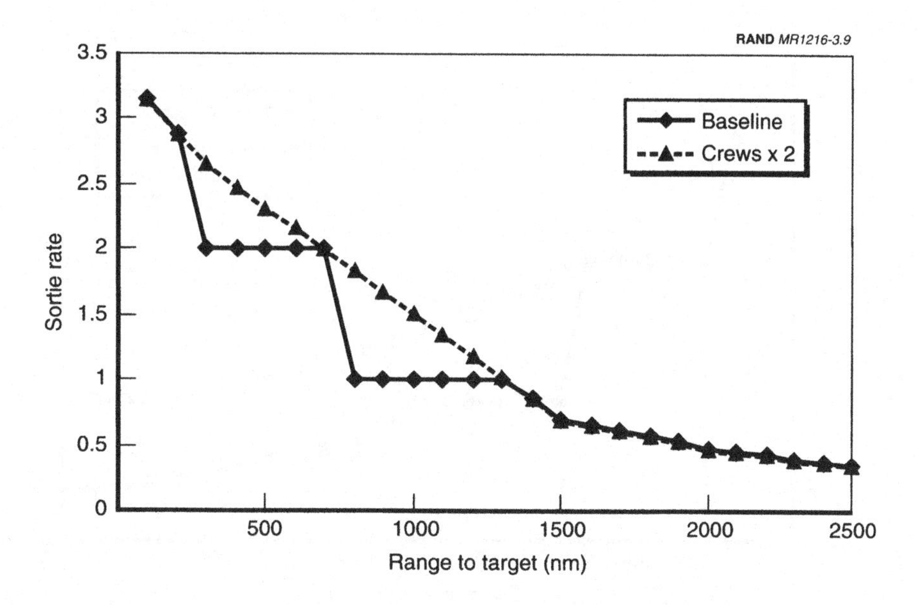

Figure 3.9—Sortie Rate vs. Range-to-Target and Crew Ratio

roster of aircrew. While tanker capacity is the binding constraint at long ranges, it lies not very far below the maximum capability of a force with only an effective 1.0 crew ratio. Therefore, one key to improving the AETF's combat capability at longer ranges would appear to lie in increasing both tanker and aircrew availability.

Increasing Crew Ratios *and* Tanker Support

Figure 3.10 shows the effect of simultaneously doubling the effective crew ratio *and* providing all the refueling capacity that the AETF's deployed fighter component can use. It shows that for all ranges beyond about 300 nm—that is, all ranges for which refueling is a factor—a significant gain is achieved in sortie generation. Indeed, the

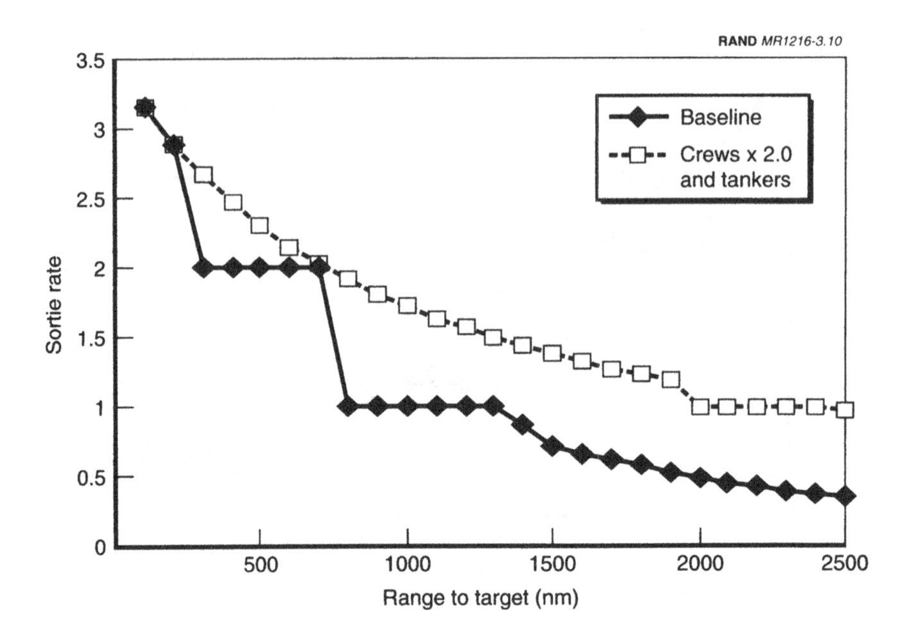

Figure 3.10—Sortie Rate vs. Distance and Crew Ratio/Tanker Support

force is operating at or near the estimated aircraft maintenance constraint out to a range of 2500 nm.[33]

These changes buy back much of the combat power that was lost when the aircraft were based at the more remote airfields. Figure 3.11 shows that the fighters are now generating about 84 strike sorties each day compared to some 88 when they are based closer in and about 51 when distantly based and unaugmented.

This added capability, however, comes at a significant cost in fuel use. The chart shows that when based close in, the deployed force is

[33]Although Air Force regulations require only 12 hours of crew rest after a mission, our model assumes that crews require 24 hours off after flying a mission of 2000 nm or greater radius. In addition, it is well worth noting that missions beyond about 2000–2200 nm to target would require a waiver of the USAF's 12-hour duty-day restriction.

Although we took maintenance capability as a given in this analysis, augmenting maintenance manning and supply could further increase long-range operational capability once tanker and aircrew constraints are relaxed.

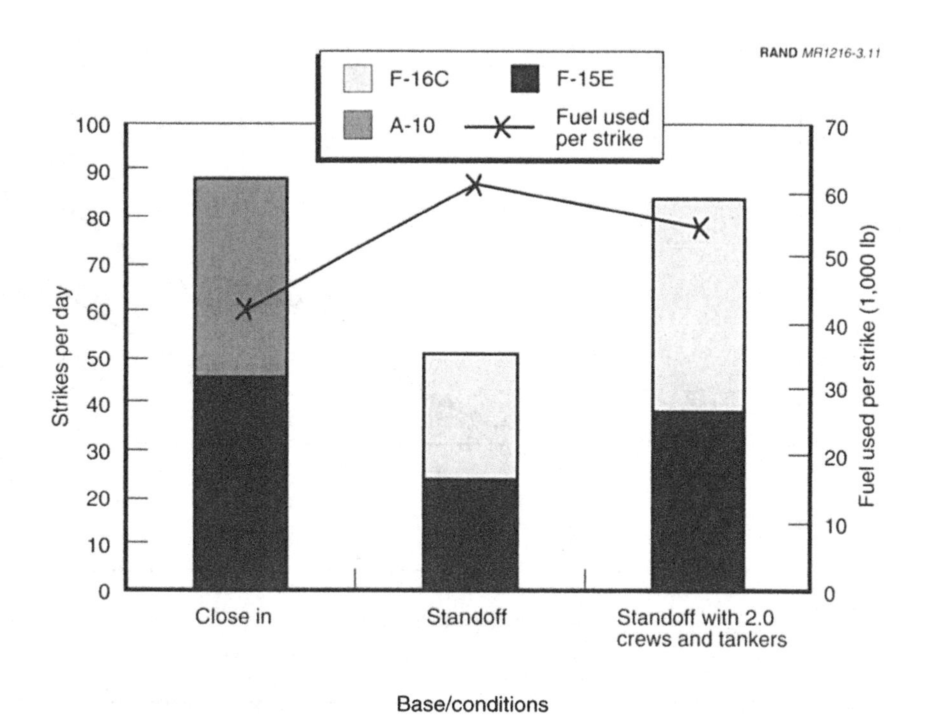

Figure 3.11—Impact of Adding Aircrew and Tankers on Sorties and Fuel Use

using about as much fuel per strike sortie as the deployed U.S. air forces did in the 1991 Gulf War—about 42,000 lb. When the force is flying with additional crews and tankers, that figure increases by about 28 percent to about 54,000 lb per strike sortie.[34] Our data did not permit us to evaluate whether our selected beddown locations have sufficient pumping capacity to support this pace of fuel usage.

Figure 3.12 shows the total tankers required to support fighter operations as a function of fighter crew ratio and range to target. The number of refueling aircraft necessary goes up dramatically as the distance between the fighters' beddown location and their targets in-

[34]Of this total, the tankers themselves burn more than one-third—22,000 lb—to deliver the remainder to the fighters.

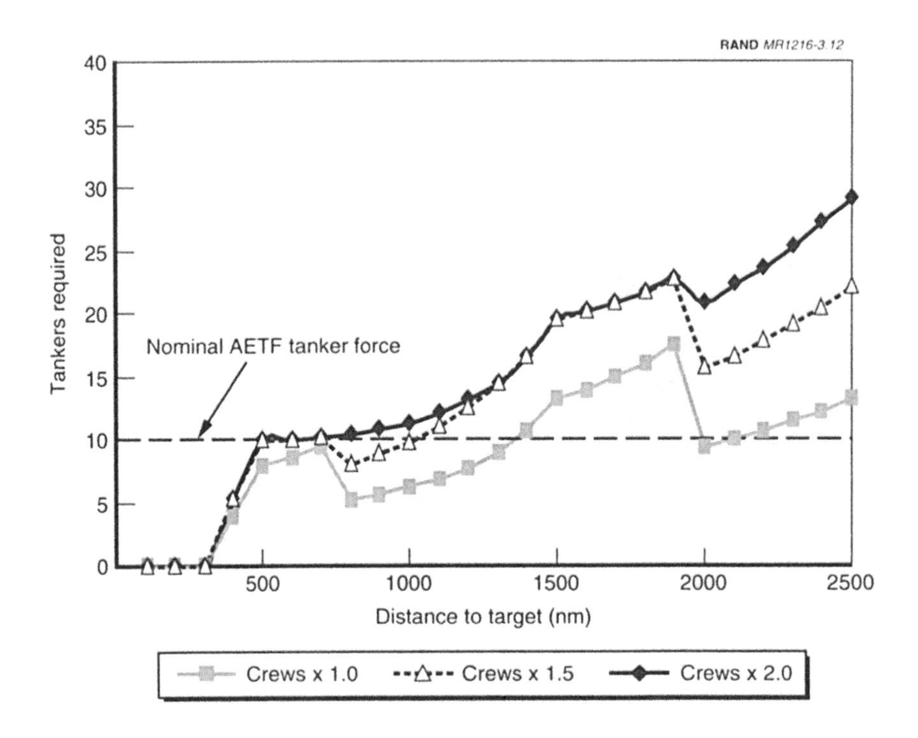

Figure 3.12—Tanker Force Required as a Function of Crew Ratio and Range

creases. Even at 1500 nm, however, the total number of tankers needed is only 15, just five more than the ten that are part of the nominal AETF.[35]

SUMMARY AND CONCLUSIONS

This chapter has attempted to cover a great deal of ground. We wish here to recap the key points that have been made:

---

[35]The figure assumes that all tanker forces, like our typical AETF tanker force, are composed of about 60 percent KC-135s and 40 percent KC-10s. However, only some 10 percent of the USAF's more than 600 tankers are KC-10s. Therefore, it may not be possible to keep the KC-10/KC-135 ratio at 4:6 as the size of the tanker force assigned to the AETF increases. To the extent that the smaller KC-135s are substituted for KC-10s, more tankers will be required than the calculations presented here indicate.

- Based strictly on aircraft operating characteristics, there are a number of locations suitable for expeditionary air force deployment in Southwest Asia. However, geography, political factors, adversary threat capability, and commanders' willingness to accept risk could interact to limit and narrow choices, especially for large, vulnerable support aircraft.
- This narrowing of options could lead the USAF to deploy assets to fields far from their intended targets, requiring long missions that are hard on fighter crews, and consuming large quantities of fuel.
- The combat capabilities of an AETF can decrease dramatically when the aircraft are forced to base at increasing distances from their intended operational areas.
- In the short run, modest increases in fighter crew ratios and tanker support could allow the typical package of USAF fighters to operate with about the same effectiveness from ranges of 1000–1500 nm to target as they can from about 500 nm.[36]

Increasing the AETF's ability to operate from distant bases would allow U.S. commanders options for hedging against a variety of factors, including increasing enemy missile capability, uncooperative regional partners, or inconvenient theater geography. In general, the farther an AETF can project effective combat power, the more options commanders will have for dealing with any and all of these factors.

In the long run, if expeditionary operations are truly the future mode of USAF employment, it may be desirable to acquire a fleet of combat aircraft that is better suited to the demands of long-range operations. The current mix of aircraft, designed during the Cold War, is optimized to fight a relatively short-range air campaign in Central Eu-

---

[36]At longer ranges, it may also be desirable to replace the A-10s in a standard AETF with faster jets, such as F-16s.

With an effective crew ratio of 2.0, operating from these distances would require fighter crews to fly a 4.3-to-6.5-hour mission about once every other day. This is an OPTEMPO similar to that sustained by F-117 pilots for 43 days during Operation *Desert Storm.* Many F-117 pilots used stimulants ("go pills") to remain alert toward the end of their long flights so even with additional crews USAF fighter units might have to resort to such measures again for extended operations from long range.

rope or on the Korean peninsula. The next generation of USAF fighter and attack aircraft, the F-22 and F-35 JSF, will likely have about the same range as current systems, making them no more capable of conducting extended-range operations without heavy tanker support. The USAF may want to consider whether improving its flexibility and capability for challenging future expeditionary operations makes it worthwhile to consider a new generation of longer-range, higher-speed combat aircraft.[37]

Deterring, fighting, and winning the nation's wars is the primary purpose of the U.S. armed forces. However, the military has long been involved in a range of other activities, including disaster relief both within and outside the United States, peacekeeping missions, and the like. The pace of these MOOTWs seems to have increased significantly in the 1990s, and there is little reason to believe that the demand for such undertakings is going to diminish in the near future. Proper access and basing are as critical to these operations as to the kinds of war-fighting campaigns that have been the subject of this chapter. In the next chapter, we will discuss the kinds of demands that could arise in the challenging MOOTWs that could characterize the first decade of the new century.

---

[37]See Stillion and Orletsky (1999), Appendix C, for a discussion of one concept for such a platform.

Chapter Four

ACCESS IN OPERATIONS OTHER THAN WAR: RAPID, SUSTAINABLE DEPLOYMENTS TO REMOTE LOCATIONS

INTRODUCTION

An ongoing civilian humanitarian relief operation in central Africa has gotten caught in the midst of ethnic civil war brought about by Tutsi refugees fleeing the Democratic Republic of Congo (DRC). The refugees' plight has elicited a relief operation of food and supplies from the industrialized countries, coordinated by the United Nations High Commissioner for Refugees (UNHCR). Over the course of several months, this has become a major undertaking involving 500 Western aid workers and several thousand Westerners and locals employed by nongovernmental organizations (NGOs) assisting hundreds of thousands of refugees in feeding centers. Meanwhile, emboldened by their victory in the DRC, Hutu rebels in Burundi have been increasingly active. They have seized control over several towns and are moving in closer to the capital, Bujumbura. The minority Tutsi-dominated government has responded with brutal, indiscriminate repression, causing many Hutus to flee to the Congo and Tanzania. In a vicious spiral of increasing violence, both sides are committing massacres. Militants are actively recruiting in refugee camps, and they dominate what little structure these camps have. The killing appears to be going out of control, Western citizens are trapped in the midst of the escalating violence, and perhaps a million people are in danger of starvation or epidemic disease. "Another Rwanda" appears to be in the making, and the call goes out for massive and rapid intervention to stop the killing, protect Western citizens, and provide massive

> quantities of relief to hundreds of thousands of refugees and internally displaced persons (IDPs).

The hypothetical passage above could have come out of today's headlines and may yet appear in tomorrow's. In the previous chapter, we saw how access and basing issues could affect the USAF's ability to deter or prosecute a major theater war. But not all future challenges will be of that ilk. Many—indeed, most—overseas operations will likely be of the kind often referred to as military operations other than war. This class of military action—which includes humanitarian aid, peace operations, crisis response, enforcing sanctions, and even military intervention in less developed countries—will almost certainly dominate the day-to-day agenda of operators and planners alike, as it has for most of the past decade.

Although MOOTWs have been steadily increasing in frequency, the Department of Defense has been inclined to view them as lesser-included cases for force planning and basing arrangements.[1] This has been a reasonable response given two assumptions: First, in the past policymakers have paid only sporadic and limited attention to the implications of likely future crises in areas that are of minimal direct strategic importance to the United States.[2] Second, recent experiences in Latin America and sub-Saharan Africa have reinforced the prevailing perception that the vast majority of MOOTWs can be accomplished by ad hoc deployments of several C-130s and a handful of personnel from bases in Germany or the continental United States (CONUS).[3]

---

[1]Builder and Karasik (1995), p. 4.

[2]Nowhere has this been more clearly reflected than in sub-Saharan Africa. Events in Africa have received little attention owing to a perception that there is little political interest in potentially costly interventions in an area so far from U.S. borders and of limited strategic interest. Then–Secretary of Defense William Perry, commenting on U.S. operations in Rwanda, argued: "Our concerns [in Africa] are primarily moral and symbolic. That does not make them less relevant, but it does help define the limits of feasibility. Our objective should be to ameliorate catastrophe and meet basic human needs. As soon as the humanitarian operation is up and running effectively, we want to get out and turn things over to relief agencies." See Schmitt (1994).

[3]For example, Operation *Noble Response* required two Marine KC-130s and 34 Marines to deliver two million pounds of food assistance to Kenya in January–March 1998.

However, one can conceive of a plausible and much more stressful scenario in which the United States might be involved—a *major* peacekeeping and humanitarian mission in a remote area with relatively little intelligence or logistical support and limited infrastructure. Without adequate planning, this type of mission could present a daunting challenge in terms of both rapid deployment and manageable sustainment. Keying off recent experiences in Somalia (1992–1993) and Rwanda (1994), we created the above scenario for a peacekeeping and humanitarian mission centered in Burundi but spilling into the entire Great Lakes region of Central Africa, an area that is remote and largely characterized by bare bases and complex politics. This scenario serves to illustrate the types of challenges that Air Force planners can expect to face in a truly complex MOOTW.

THE CHALLENGE OF COMPLEX OPERATIONS OTHER THAN WAR

Postmortems on the United States' and Western response to the 1994 events in Rwanda have suggested that initially the United States and its partners grossly underestimated the peril that the average Rwandan citizen faced.[4] Subsequently, then-President Clinton and other members of his administration publicly pledged that the world would not permit a repeat of such a scenario—that there will not be "another Rwanda."[5]

Thus, while Africa has been and continues to be viewed as peripheral to any direct or vital Western strategic interest, a credible threat of another massive ethnic conflict, particularly in an area of such geographic proximity to Rwanda, is likely to attract the attention of U.S. policymakers. Under such circumstances, the call might well go out to the military services to lead an intervention to stop the killing and facilitate the provision of basic humanitarian assistance.

While on the surface this appears to be a relatively limited mission, analysis of current regional dynamics, previous experience in intervening in ethnic conflicts, and historical experiences in Rwanda and Somalia raise serious logistical and operational questions. Tasked

---

[4]Feil (1998). See also Gourevitch (1998).

[5]Schutz (1998).

with significant participation in a MOOTW in Central Africa, for example, the USAF would face a tradeoff between the political impossibility of allowing another massacre to occur and the operational obstacles to preventing it.

Sizing the Force: Lessons from Somalia and Rwanda

The size of the forces required in any scenario will, of course, vary from circumstance to circumstance but is also likely to vary across time within a given scenario. An initial intervention force may need to be small, fast-moving both strategically (to arrive at the scene in a timely manner) and tactically (to make its presence felt where and when needed), and highly capable. A larger force may subsequently be needed to handle relief and rebuilding over the longer term.

The situation we are considering is one in which confusion is rampant, infrastructure is sparse, and no friendly forces are on the ground. Under such circumstances, accomplishing the tasks set forth—to head off or end the killings and to secure the distribution of relief supplies and medical care—will likely require a sizable force.

Although the military capabilities of potential adversaries in these missions are limited, one of the lessons from interventions such as Operation *Restore Hope* in Somalia is that preponderant force serves an important purpose.[6] As one analyst wrote, "During conflict, parties regard humanitarian assistance as a means to enhance their power and degrade their adversaries. Only strong military force can prevent them from diverting and misappropriating assistance."[7] In the five months that the United States participated in *Operation Restore Hope*, it deployed forces that included two brigades of the Army's 10th Mountain Division as well as extensive divisional and nondivisional support, a Marine Expeditionary Brigade, a carrier battle group, an amphibious-ready group, and a Maritime Prepositioning Squadron. Coalition forces reached their peak in January

[6]The Somalia experience also serves notice that even poorly equipped paramilitary forces can be formidable adversaries under the right—or wrong—circumstances. It was, after all, a lucky rocket-propelled grenade that brought down a U.S. Army helicopter in Mogadishu and touched off "Bloody Sunday." See Bowden (1999).

[7]Pirnie (1998), p. 63.

1992, when personnel numbered more than 38,000, of whom 25,426 were U.S. troops.[8]

A thought-provoking paper by Scott Feil argues that a properly configured and enhanced brigade of the U.S. 101st Airborne Division (Air Assault) could have forestalled the Rwandan genocide had it been rapidly deployed in early April 1994.[9] The force envisioned by Feil would have consisted of nearly 5800 troops and more than 70 helicopters.

Anticipating Demand for Airlift

In a complex MOOTW, the USAF could be called upon to deploy its own assets to bare bases in the area of operations, to provide airlift for U.S. and regional-coalition ground troops, to support noncombatant evacuation operations (NEO), and to transport some quantity of humanitarian supplies. What might these requirements be?

The first response to a regional crisis is likely to be the search for a local solution. In Africa, for example, there have been several proposals by African states, the United States, and France to establish and train a force of 5000–10,000 troops for peace operations on the continent.[10] At the insistence of South Africa and Kenya, these forces would be African-led, African-manned, and capable of both peacekeeping and crisis response. Setting aside concerns regarding the implementation of this training program, these troops would lack the airlift to arrive on site in a timely manner and would be so limited in size that they could constitute only a small proportion of the ground troops required for a truly difficult contingency.

Table 4.1 indicates the number of airlift sorties required to move various U.S. ground force units. If the area of operations is inland and remote, forces either would have to be airlifted in or would have to conduct a potentially arduous road march from an available sea port. Given the poor transportation infrastructure in much of sub-

---

[8]Hirsh and Oakley (1996).

[9]Feil (1998).

[10]Christian (1998).

Table 4.1

Airlift Required to Move U.S. Army Units

	Airlift Sorties Required		
Unit Type	C-141	C-5	Approx. C-17 Equivalents
Airborne division	1125	34	488
Air assault division	1330	161	711
Light infantry division	811	39	369
Light armored cavalry regiment	477	15	208
Separate infantry brigade	326	10	142
Separate mechanized brigade	418	241	436
Theater support assets	600	105	357

SOURCE: Military Traffic Management Command, Transportation Engineering Agency, *Deployment Planning Guide*, MTMCTEA 97-700-5, July 1997. Assumes sufficient civil airlift or charter available to move most personnel.

Saharan Africa, just getting to the operational area could prove extremely difficult under such conditions.

While contract carriers would likely be employed to move the bulk of needed humanitarian aid, some urgent or specialized cargo might need to be transported by the USAF. For example, U.S. military airlift might be used to deploy water purification equipment or to bring in initial supplies of food and medicine to sustain refugees and IDPs until commercial services could be set up to take over this task.

Minimum water needs vary with each situation but increase markedly with raised air temperature and physical activity. Table 4.2 lists some rough factors used for humanitarian relief planning. A U.S. Army water purification detachment, which deploys in six C-141 and four C-5 sorties, can produce 30,000 gallons of water per hour. This could supply about 70,000 people (assuming approximately 10 gallons per person per day for personal and feeding-center consumption) if the water could be distributed in a timely and effective manner. In most cases, establishing a reliable distribution process will represent the biggest challenge to providing refugees/IDPs with adequate water, and providing the equipment necessary to set up such a system—such as tanker trucks—could prove an additional burden on airlift.

Table 4.2

Water Requirements

Requirement	Water Needed (liters per person per day)
Drinking, food preparation, cleanup	3–4
Personal hygiene	2–3
Laundry	6–7
Feeding centers	20–30
Health centers	40–60

SOURCES: U.S. Agency for International Development, *Field Operations Guide for Disaster Assessment and Response*, version 3.0, available at http://www.usaid.gov/ofda/fog/, August 1998; The Sphere Project, *The Humanitarian Charter and Minimum Standards in Disaster Response*, available from http://www.sphereproject.org/handbook_index.htm, n.d.

In terms of food, the United States Agency for International Development (USAID) recommends a ration based primarily on cereals, pulses such as lentils and beans, and vegetable oils. Such a menu can deliver a reasonably balanced, 2100-calorie "survival diet" that weighs only about 540 grams. The U.S. military has developed a humanitarian daily ration (HDR) that provides "full day's sustenance to a moderately malnourished individual."[11] To make it palatable across the widest possible range of cultures, the HDR contains no meat or animal products and no alcohol.[12] An HDR weighs in at about a kilogram, and 48 cases of ten HDRs each can fit onto a standard cargo pallet. A C-17 can carry 18 pallets, so a single sortie could lift 8640 HDRs.[13] Multiple airlift missions, then, might be needed just to provide an adequate initial stockpile of food.[14]

---

[11]See entry for Humanitarian Daily Ration on the Defense Supply Center-Philadelphia Web site: http://www.dscp.dla.mil/subs/rations/hdr.htm, browsed June 2000. We are grateful to Paul Killingsworth for providing us with this information.

[12]Even the moist towelette provided for cleanup is specified as alcohol-free.

[13]A seven-day supply for 20,000 people, or 140,000 HDRs, would require about 16 C-17 sorties. Reports from the initial stages of Operation *Enduring Freedom* indicate that even larger numbers of HDRs can be air-dropped from C-17s using the tri-wall delivery system. Two C-17s are reported to have delivered about 35,000 HDRs over northern and eastern Afghanistan on October 8, 2001. See Mitchell and Fidler (2001).

[14]Malnourished children, pregnant or lactating women, the elderly, and the ill often require a supplementary ration. We do not have any data available from which to calculate the airlift requirements for providing these to a sizable refugee population.

In absolute size, these demands for airlift—a dozen or so sorties to deploy water purification equipment and another 10 to 20 for foodstuffs—hardly compare to the requirements for a major deployment. However, airlift resources have proven to be heavily tasked on a day-to-day basis over the past ten years, and there is little to suggest that this situation will soon change for the better. A sudden need for 30 to 40 immediate strategic mobility missions, in addition to however many missions are needed to deploy U.S. and other forces, could significantly stress the system. Add to this the limitations of available basing, a possible lack of fuel at the receiving end, and poor infrastructure to support the onward movement of supplies that have been flown in, and the potential for delay—perhaps with tragic consequences—appears very real.

Timing for Deployment: Lessons from Rwanda

The deployment of 28,000 U.S. troops to Somalia for Operation *Restore Hope* required a long lead time for the Air Force to establish strategic air bridges to U.S. bases and other facilities worldwide.[15] However, the experience of Rwanda suggests that an intervention to halt a genocide may require much more rapid response. The violence in Rwanda was a planned and systematic massacre conducted by lightly armed militias and civilians occasionally assisted by the *gendarmerie,* or army. Within hours of the death of President Habyarimana on April 6, 1994, violence had broken out. By May 5, a month later, Hutu-controlled radio proclaimed a "cleanup day": "The final elimination of all Tutsis in Kigali."[16] Within three months between half a million and 800,000 Rwandans, most of them ethnic Tutsi, were dead, another half million were displaced within Rwanda, and more than two million had fled to surrounding countries. Clearly, a very prompt deployment would have been needed to prevent any substantial portion of the violence.

Demand for these supplementary foods is difficult to predict, but a significant proportion—perhaps 20 percent or more—of refugees from an ethnic conflict in a less developed country may require them.

[15]Allard (1995), p. 41.

[16]Gourevitch (1998), p. 134.

As mentioned earlier, Feil argues—and General Roméo Dallaire, commander of UN forces in Rwanda in 1994, concurs—that a modern force of 5000 troops drawn primarily from one country, willing to take combat risk, and sent within the first three weeks could have significantly altered the outcome in Rwanda. This force would have been tasked with seizing, at one time, key objectives all over the country and would thus have stemmed the violence in and around the capital, prevented its spread to the countryside, and created conditions conducive to a cessation of civil war.[17] How rapidly could such a force have moved into Kigali?[18]

We used standard U.S. Army and USAF reference materials to assess the airlift that would be necessary to deploy a task-organized brigade of the 101st Airborne Division (Air Assault) and to estimate how quickly this brigade could move into Central Africa.[19] Our results suggest that approximately 297 C-141 and 60 C-17 sorties would be necessary to move a force consisting of

- Five air assault infantry battalions
- One assault aviation battalion of UH-60 helicopters
- One medium-lift helicopter battalion of CH-47s
- One AH-64 attack helicopter battalion
- One forward support battalion
- One military intelligence company
- One signals company
- One military police (MP) company

---

[17]Feil (1998).

[18]Whether a rapid force deployment would have been decisive in curtailing the Rwandan genocide is the subject of some debate. For a contrarian view, see Kuperman (2000), pp. 94–118.

[19]Military Traffic Management Command (1997) provides an estimate of the airlift required to move the individual elements. AFPAM 10-1403, *Air Mobility Planning Factors* (U.S. Air Force Air Mobility Command [1997]), allowed us to use those requirements as a basis for time-to-close calculations.

- One chemical warfare defense company
- One headquarters company.

In addition, approximately nine charter flights of Boeing 747–class aircraft would be needed to move some 3000 personnel who could not be accommodated on the military transports.[20]

How rapidly these aircraft could move their cargo into the area of operations would depend on the number and quality of the aerial ports of debarkation (APODs) available there. This timeline is especially sensitive to three factors:

- The number of transports that can be on the ground loading or unloading at any given time, referred to as the "maximum on ground" (MOG).
- The number of hours each day the APODs are operational.
- The number of airlift aircraft committed to the mission.

Our calculations show that each constraint can dominate the result under certain circumstances. For example, if the MOG were 1 (that is, if only a single transport could be loading or unloading at any given time) and the APOD were capable of only daylight operations—a situation that could arise if all traffic had to move through a single underdeveloped international airport—it would take about 40 days to close the force. Under such conditions, throughput on the receiving end would simply be too limited to permit a faster deployment regardless of how many airlifters are available. Conversely, if the MOG were 3.0 and the APOD or APODs were running 24 hours a day, the number of transport aircraft would become the driver. If 40 C-17s were available—a full third of the originally planned C-17

[20]These figures assume, as does Feil, that the CH-47 battalion would self-deploy. Given that the cruising speed of the Chinook is between 120 and 140 knots and its ferry range—carrying full fuel but no payload—is 1111 nm, this could take quite some time even if the helicopters come from Europe. Aircraft data are from the U.S. Naval Institute (USNI) *Periscope* database, http://www.periscope.ucg.com/weapons/aircraft/rotary/w0004511.html (1999). Also, sustaining the force in action would require that further combat and combat service support elements be deployed. As configured, the brigade task force could probably operate for no more than seven to ten days without further support and resupply. (From conversations with retired U.S. Army officers at RAND.)

buy—the force could close in 18 days. A middle case—a MOG of 2.0, 18 hours of APOD operations per day, and 30 C-17s committed—would require 24 days to close the force, driven by APOD limitations.[21]

Using this last scenario as a not-unreasonable estimate and assuming that the first transport serial launched on April 8, one day after the organized violence began in Kigali, the brigade would have finished deploying by around the first of May. By then, tens of thousands of Tutsis would almost certainly have perished—perhaps more if knowledge of the impending Western intervention motivated the *genocidaires* to increase the pace of their work so as to be more nearly done by the time the foreign soldiers arrived.[22] While many lives might in the end have been saved, even a heroic deployment effort would likely have been "too little, too late" for many victims of the Hutu genocide.

Simply establishing the necessary infrastructure to begin such a force movement could prove difficult and time-consuming given the shortage of suitable runways or support facilities that the USAF would confront in Central Africa. Even the limited level of activity required for Operation *Support Hope* to the Congo and Rwanda in 1994 required that the USAF Air Mobility Command (AMC) deploy tanker airlift control elements (TALCEs) to Addis Ababa, Ethiopia; Entebbe, Uganda; Mombasa, Kenya; Goma, Zaire; Harare, Zimbabwe; Kigali, Rwanda; and Nairobi, Kenya.[23] The luxury of flowing most men and materials into a single well-equipped airhead as was done in Saudi Arabia during Operation *Desert Shield* simply does not exist in this part of the world.

---

[21]Note that in actual operations, APODs would likely be further stressed not only by the need to transship cargo from strategic airlifters to smaller aircraft such as C-130s but also by the limitations of the local distribution networks.

[22]That the perpetrators would have been encouraged to step up their homicidal rampage rather than be deterred by imminent Western military action might have seemed ridiculous to many in 1994. After witnessing the Serbian reaction in Kosovo to the onset of NATO's air campaign, it somehow seems more plausible.

[23]Pirnie and Francisco (1998), pp. 64–65.

Limits on Basing

What are the basing options that the Air Force would have to cultivate in order to plan and conduct a large MOOTW in a remote region such as sub-Saharan Africa? We examined suitable airfields and runways on the continent using unclassified airfield data and screening them against published planning factors for aircraft airfield restrictions.[24] Not surprisingly, options are significantly constrained by the limited infrastructure. Specifically there are only eight airfields in six countries that are suitable for operating KC-10s[25] and only 16 bases in nine countries that could handle C-17 or C-5 aircraft.[26] This sparse set of basing alternatives for large transport aircraft means that the USAF might have to set up one or more hub bases at a significant distance from the theater of operations. The last leg of the trip would then be made by theater airlift (C-130s) or via ground transport. As Figure 4.1 suggests—using Burundi as the ultimate destination—some of these residual distances could be quite large.[27]

CONCLUSIONS AND IMPLICATIONS

As we worked through our scenario for intervention in Burundi, several things became clear.

---

[24]U.S. Air Force Air Mobility Command (1997). The figures used were: 6000-ft runway length, 147-ft width, and 80 load capacity number (LCN, a metric for runway pavement strength) for a C-5; 3000-ft runway length, 90-ft width, and 94 LCN for a C-17; and 7000-ft runway length, 148-ft width, and 102 LCN for a KC-10. These are minimum landing lengths for a fully loaded aircraft and assume that the transport will be taking off mostly empty or at least substantially below its maximum possible weight. If the airlifters were required to fly out more fully loaded, runway requirements would be stricter, and many if not most of these fields might wind up being unsuitable. For political reasons we did not include airfields in Libya or Algeria in our survey.

[25]These include Burkina Faso, Burundi, Cameroon, Egypt, Malawi, and Nigeria.

[26]These include Burundi, Kenya, Egypt, Uganda, Nigeria, South Africa, Malawi, Burkina Faso, and Cameroon.

[27]They could also be quite short; as Paul Killingsworth points out, in the Rwanda crisis the hub was established as Entebbe Airport in Uganda, just 200 miles from Kigali. As with most factors we have dealt with in this analysis, the specifics are unpredictable, which again militates in favor of maximal flexibility in USAF planning and operations.

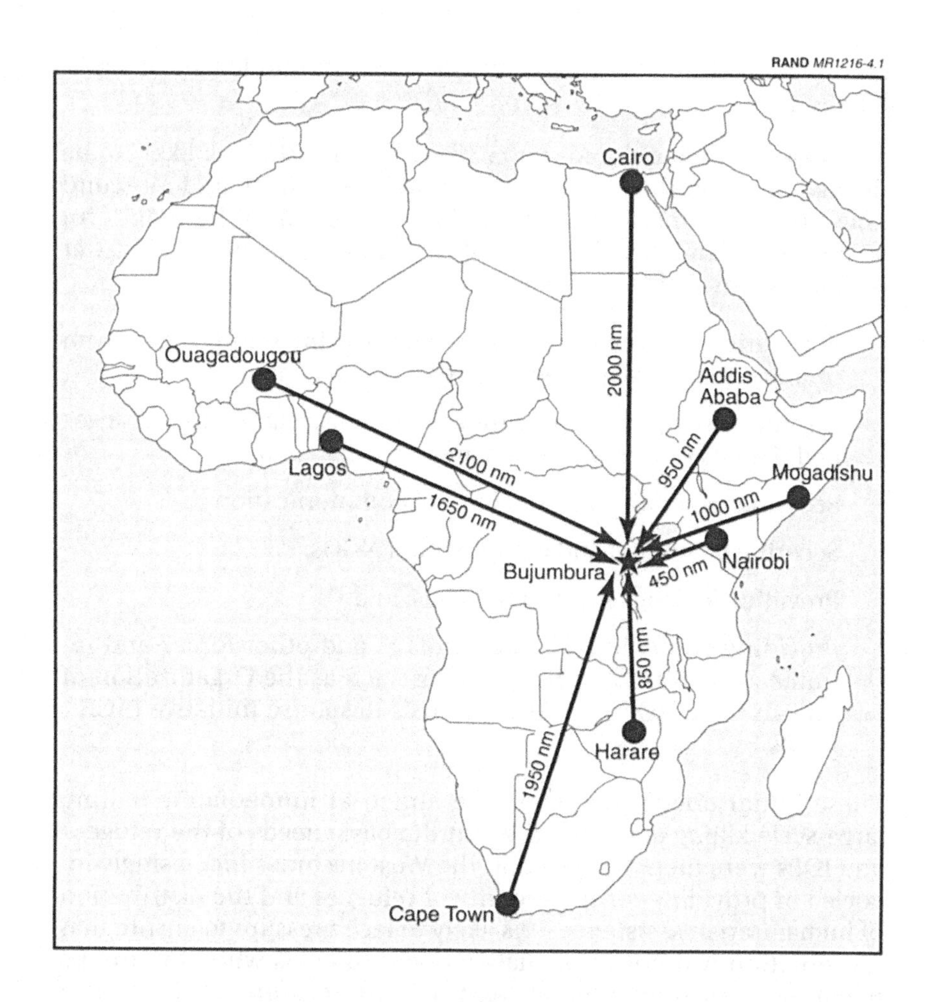

Figure 4.1—Sample Distances from Possible Airlift Hubs to Burundi

The Likelihood of "Mission Creep"

First, halting genocide and providing for basic human needs (food, water, and shelter) may be neither a limited nor a simple mission. If Burundi descended into a civil war such as that which Rwanda experienced, simply ensuring that food was reliably available throughout the country would require a significant military presence on the ground with sizable demands for Air Force lift and logistical capabilities.

The missions facing the Air Force in such a MOOTW are likely to be divided between those that have immediate priority and a second phase or tier of missions that would be essential to the long-term achievement and maintenance of the first-order goals. First-tier missions seem likely to include

- Evacuation of Western citizens, possibly including those from multiple remote sites where fighting may be ongoing;
- Ensuring free passage of humanitarian assistance for refugees and IDPs as well as those who are in border camps;
- Securing major airports and lines of communication;
- Securing personnel and equipment of NGOs;
- Providing logistics support to NGOs; and
- Providing strategic lift to deploy U.S. and other forces and intratheater airlift for regional forces such as the Organization of African States (OAS) or African Crisis Response Initiative (ACRI) forces.

These initial operations would be aimed at immediately halting large-scale killing and ensuring that the basic needs of the refugees and IDPs were met. However, as the West becomes increasingly involved in providing for the security of refugees and the distribution of humanitarian assistance, it is likely to face pressure to ensure that the situation will not immediately revert to crisis when the troops depart. To prevent this, at least some forces are likely to face additional missions. These could include

- Gaining freedom of movement and demonstrating overwhelming force to warring factions;
- Dismantling unauthorized checkpoints and suppressing banditry;
- Conducting disarmament as necessary to establish a secure environment;

- Repairing or upgrading key infrastructure to support operations; and
- Providing surveillance of the area of operations, including border camps.

The Need for Rapid Response

Second, there would be a sense of urgency in getting the forces and support elements in quickly, ideally within two to three weeks from the moment the crisis heats up. This timetable might well come to grief given the realities of preparing forces to move and actually moving them, particularly in the absence of adequate advance planning. These problems would also be exacerbated in an area such as Central Africa, where the infrastructure to support a major airlift is limited.

Based on previous interventions of similar scale, USAF force elements that could be called on for deployment include

- Strategic airlift (C-5, C-17, C-141);
- Intratheater-lift assets (C-130);
- Air-refueling aircraft (KC-135, KC-10);
- Reconnaissance elements (Joint Surveillance and Target Attack Radar System [JSTARS], etc.);
- Electronic combat aircraft (COMPASS CALL, COMMANDO SOLO);
- Special operations squadrons (AC-130, MC-130);
- Multiple TALCEs;
- Air intelligence assets;
- Airborne medical evacuation squadrons;
- Multiple aerial port units;

- Engineer units (RED HORSE); and
- Multiple security forces (SF) flights.[28]

In cases such as Burundi, where the area of operations is landlocked, an even greater strain may be placed on the Air Force.

Limited Infrastructure and Politics Play Havoc with Access

Finally, the limited infrastructure in this region—few runways and airfields capable of supporting large aircraft, limited ramp space, and a shortage of refueling facilities—could require multiple bases at a significant distance from the area of operations. This would present an intensive demand on specialized USAF units (engineers, SF, aerial port squadrons, and the like).

In the specific case of an intervention in Burundi, the Air Force would face the challenge of a limited and largely primitive infrastructure. While the airfield in Bujumbura can accommodate KC-10s, there is only one landing surface and only limited refueling capabilities are available. This makes the operational difficulties similar to those encountered in Somalia, where the Mogadishu airport was able to handle only two aircraft at a time.[29] Furthermore, the Bujumbura airport is the only one in Burundi with a paved runway, although there are two other unpaved airfields that could accommodate a C-130. Burundi has no railways and only 1000 km of paved roads.

Multiple bases can be problematic politically as well as logistically. Particularly as the timeline for operations stretches out longer and longer, political complications associated with particular basing choices are likely to become increasingly salient. Ethnic and political divisions run deep, and not just among the locals; France, for example, remains interested in former colonies such as Rwanda and would look askance at any intervention that appeared to slight its preferred party or parties. Planning even at the conceptual level for operations such as those we have described herein must take these

---

[28]Pirnie and Francisco (1998), p. 33.

[29]Allard (1995), p. 46.

cross-cutting sensitivities into account if it is not to run afoul of them.

SUMMING UP

Our work suggests that future complex MOOTWs could be highly demanding for the USAF and should probably not be dismissed as lesser-included contingencies. Instead, more planning may be called for to ensure that the USAF is both operationally and politically prepared to mount the rapid and sustainable deployments that are integral to such missions. In Africa and elsewhere, flexibility will be the key.[30] Maintaining existing strategic relationships with key actors such as Egypt and Kenya will be a vital component of ensuring adequate access, but the United States should also seek to strengthen its relationships with other potential hosts. In Africa, candidates might include South Africa, Zimbabwe, and Ethiopia, among others.

Achieving a degree of flexibility in planning and operations is one critical element of an overall strategy for ensuring that needed access and basing are available for future USAF expeditionary operations. In the next and final chapter, we will outline one such strategy.

---

[30]While this chapter has focused on Africa, similar logic and conclusions would apply in other areas of the world, such as Latin America.

Chapter Five

DEVELOPING A GLOBAL ACCESS STRATEGY FOR THE AIR FORCE

Access will remain a challenge to the U.S. military in general and to the Air Force in particular for the foreseeable future. The preceding pages tell a story that is part good news and part bad news. On the positive side

- The United States enjoys strong defense relationships with a large number of countries all around the world. This web of engagement serves to facilitate access for the USAF.
- While access has historically been an irritant on many occasions, U.S. diplomacy, flexibility, and luck have usually resulted in the availability of workarounds to enable operations.
- There are a number of countries that, in looking to improve or cement their security relations with the United States and the West, could be strong candidates for enhanced access arrangements.
- Given some modifications in manning and support, current and future USAF forces appear able to sustain a reasonably high tempo of operations at fairly long ranges from their operational areas—up to 1000–1500 nm.

The negatives are as follows:

- "Assured access" is a chimera outside U.S. territory. National sovereignty may be eroding in cyberspace, but in the "real world" of air bases and airspace, it continues to reign supreme.

- Even close allies, such as the British and Germans, have at times refused access or overflight.
- In addition to the politically driven access problems that the United States has occasionally encountered, new military threats—particularly advanced surface-to-surface missiles—may change the calculus of risk, inducing commanders to base forces farther away from the immediate combat zone.
- Access arrangements in Southwest Asia and Asia outside of Korea and Japan are limited and may prove woefully inadequate for the kinds of contingencies that could develop in those regions.
- Given current and likely future access arrangements, it could prove very difficult to project and sustain a significant amount of power into sub-Saharan Africa and Latin America south of the Equator. The former in particular appears to pose serious challenges.

In short, the USAF confronts a complex set of circumstances; what options exist for dealing with them successfully?

FIVE "PURE" STRATEGIES

We have identified five alternative approaches toward managing access and basing in the future.[1] They are

- Expand the number of overseas main operating bases (MOBs) to increase the likelihood that forces will be present where and when they are needed.
- Identify one or more "reliable" allies in each region of the world and count on them to cooperate when asked to do so.
- Proliferate security agreements and alliances to broaden the set of potential partners in any given contingency.
- Negotiate and secure long-term extraterritorial access to bases, as was done with Diego Garcia.

[1]A sixth strategy is hinted at in the first sentence of this report: imperial conquest as in the British Empire of old. Suffice it to say that none of the authors is at risk of losing sleep over eliminating this *a priori* as a viable option.

- Rely on extended-range operations from U.S. territory.

We believe that each of these strategies is insufficient in and of itself to ensure adequate access. We will briefly discuss each in turn.

Expand Overseas MOBs

The proliferation of a permanent presence overseas has a historical pedigree: USAF forces were at one time stationed at dozens of locations around the world. In the wake of the Cold War, that base structure has been substantially reduced. Why not rebuild a larger and more robust array of permanent overseas MOBs to support the USAF's power projection mission?

At least three serious objections can be raised to this approach:

- There would appear to be no popular constituencies, either domestic or foreign, for such an expansion.
- Unless host countries pick up all or part of the tab, foreign MOBs are expensive propositions. Freeing up money to build or reopen these facilities would thus be extremely difficult.
- Having forces stationed on another country's territory does not in itself guarantee that they can be used however and whenever they are desired. Spain, Saudi Arabia, Turkey, and others have demonstrated this repeatedly over the past quarter century.

Rely on the Reliable

Great Britain has proven to be a particularly stalwart friend to the United States—for example, by enabling the 1986 raid on Libya. Along with Turkey, Britain is the only other country that shared the burden of policing the no-fly zones in Iraq. Might the United States perhaps identify one or more "Britains" in other parts of the world whose reliability would be such that they would rarely if ever be uncooperative? Unfortunately, our analysis suggests that this would not be easy.

First, candidates are few and far between. Britain and the United States have, after all, enjoyed a mutually beneficial "special relationship" since the 1940s. It began with Lend-Lease, was solidified through the war against Hitler and with British participation in the Manhattan Project, and set was on firm ground with continued cooperation on postwar nuclear matters. Moreover, the United States has a strong cultural attachment to and affinity for Britain that is deeply rooted in both countries' common history.[2] Looking around the world, it is difficult—indeed, one is tempted to say impossible—to find another country that shares a similar range and depth of connection with and similarity of perspective to the United States. This is especially the case in Asia and the greater Middle East—the regions where access promises to be especially problematic in the near term.[3]

It also bears repeating that even "reliable" Britain has at times asserted itself by refusing to cooperate with the United States. London's failure to support the Operation *Nickel Grass* airlift to Israel in 1973 is probably the most notable example.

To be sure, this is not to say that the United States should not try to nurture close and robust relationships with other countries. It would be imprudent, however, to rest an overall access strategy on this single leg.

Expand Security Agreements and Alliances

Another option would be to greatly expand the existing network of alliances and other security arrangements that bind other countries

---

[2]One author can recall a convivial evening in a Royal Air Force mess where, after several pleasant hours of conversation and toasting one another's well-being, he found himself profusely thanking Her Majesty's forces for "wearing those lovely red coats back in '76." He doubts that he would have gotten the same amused reaction had he made a parallel remark in a *Luftwaffe* officers' club, for example.

[3]Israel might represent a plausible candidate for a "special relationship." As discussed earlier, however, Israel's somewhat shadowy status among its neighbors could impose great limitations on its utility as a point of access to the region. Should these circumstances change for the better, this assessment could change as well.

Australia may appear to be a possible "England" in the Western Pacific. However, Canberra's regional and global perspectives are not identical to those of the United States and a significant portion of its people are likely to oppose greatly expanded defense ties with the United States. Furthermore, Australia's location makes it less than ideally suited to support USAF operations outside its immediate Southeast Asian vicinity.

to the United States and vice versa. Indeed, NATO's recent expansion and the success of the PfP program has in fact opened new doors to USAF access.[4] Two points must be made in this context, however.

First, as with the idea of expanding the number of USAF overseas MOBs, it is difficult to identify the political constituencies that would support a wide-ranging extension of U.S. alliance guarantees. Domestically, support for NATO expansion may be unique, based more on post–Cold-War goodwill and public familiarity with the Atlantic alliance's long-time role in U.S. security than on any desire to see the American security umbrella more broadly spread. And while there is little doubt that America will remain an engaged and active power on the international scene, the persistence of the isolationist siren song within the national political debate may indicate that these may not be the most propitious times to advocate such an expansion.[5]

Second, as was suggested in Chapter Two, much of the payoff in terms of cooperation from enhanced security arrangements may come during the courtship as opposed to the marriage. A desire for improved relations with the United States may motivate a partner to be more cooperative than it will be when, secure in its status, those improvements are cast in stone.

"Rent-a-Rock"

The value of Diego Garcia in supporting the U.S. position in the Persian Gulf leads one to question whether there might be opportunities to make similar arrangements elsewhere in the world. To help improve U.S. access in the area around Taiwan, for example, might it be possible to lease from the Philippine government one of the many desolate, uninhabited islands in the archipelago and build an MOB there? This is an intriguing and potentially powerful idea.

Of course, only extraordinary circumstances typically induce a country to cede sovereignty over part of its territory; Britain granted

---

[4]As witness Hungary's cooperation with NATO during Operation *Allied Force*.

[5]The political consensus on U.S. overseas involvement may be in for a change in the wake of September 2001 and the subsequent military operations in Afghanistan. What those changes may be, and how enduring, cannot be ascertained at this writing.

the lease on Diego Garcia only in the wake of World War II and in the context of its historic Cold War withdrawal from "east of Suez." It can certainly be imagined, however, that some set of incentives might prompt Manila, for example, to agree to a similar arrangement with the United States. Filipino perceptions of rising hostility from Beijing, for example, could drive the Philippines to pay a high price for U.S. protection. The idea should therefore not be dismissed out of hand. There are, however, at least two reasons why this is not a complete solution to future USAF access needs.

First, these arrangements are rare indeed. Although the United States enjoys such a status at Diego Garcia and Guantanamo Bay in Cuba, the first was acquired from a close friend that no longer needed it and the second was a remnant of the colonial past.[6] To assume that Washington will be able to acquire such privileges anywhere else, let alone at multiple locations, would be foolish.

Second, we would expect that only uninhabited locales could even come under discussion as candidates for such an arrangement. And such places are typically uninhabited for a good reason, such as a pestilential climate, lack of livable real estate, or an absence of fresh water—conditions that would also present difficulties in establishing a major military installation. To be sure, none of these conditions is necessarily prohibitive; swamps can be drained, mountains flattened, and salt water made fresh through the sufficient application of ingenuity and cash. However, the upfront costs of such undertakings are likely to be very high, and the reallocation of resources within DoD to provide for them would be extremely painful.[7]

Project Power from U.S. Territory

A final option is to reduce reliance on overseas access by resorting increasingly to employing airpower from sovereign U.S. territory. The success of long-range bomber raids from bases in CONUS—

---

[6]A third, the Panama Canal Zone, was likewise a hangover from empire and passed into history within weeks of this writing.

[7]Costs are also a major factor militating against a higher-tech variant of this approach: the construction of large floating air bases. Another strike against such platforms is that, unlike England, islets, and atolls, they lack inherent unsinkability, making them potentially lucrative targets for a capable adversary.

B-52s carrying cruise missiles from Louisiana to Iraq and B-2s attacking Serbian targets from Missouri—lends this idea credibility. Moreover, the improving conventional capabilities of the USAF's heavy bomber fleet clearly earmark these aircraft for a more prominent role in future conflicts. Two factors, however, will limit the extent to which these sorts of operations can—at least in the near- to midterm—dramatically reduce the need for overseas access across all contingencies.

Sheer weight of numbers is the first factor. The USAF currently fields more than 2100 fighter and attack aircraft in comparison to some 152 bombers, and it plans no further procurement of long-range strike platforms for at least 20 years.[8] Thus, more than 90 percent of the Air Force's combat aircraft cannot and will not be able to operate effectively from U.S. territory in any but the most exceptional scenarios.

This quantitative difference looms even larger when we account for the productivity difference between a bomber based in CONUS and a fighter that is in theater. Heavy bombers flying 30- to 40-hour CONUS-to-CONUS missions must obviously generate less than one sortie per aircraft per day. In fact, for analytic purposes, it is typically assumed that a realistic sortie rate may be one every two or three days, and this appears broadly consistent with what has been achieved thus far in practice. An F-15E, on the other hand, can achieve an average of between 1.5 and 2 sorties per day when based within 1200 nm or so of its targets.[9] And although the bomber's heavy payload makes up somewhat for the disparity in sortie rates, the limited number of bombers available—in comparison to the number of fighters and attack aircraft—further reduces the heavy force's relative impact, as shown by the illustrative numbers in Table 5.1.[10]

---

[8]USAF force numbers as of September 2000 from "Equipment" in *AIR FORCE Magazine*, May 2001, p. 55.

[9]See Figure 3.10.

[10]This rough comparison ignores a host of operationally important factors, not the least of which is the value of the B-2's low-observable configuration. Nonetheless, it does, we believe, present a reasonably valid comparison of capabilities along one important dimension. Employing bombers other than the B-2 will obviously increase the amount of firepower available from CONUS bases, although neither the B-52 nor the B-1 have the same ability to operate and survive in a high-threat environment as the B-2.

Table 5.1

Illustrative Comparison of Weapon-Delivery Potential

Aircraft	Payload	Daily Sortie Rate	Weapons Delivered per Day	Weapons Delivered in Ten Days
1 × F-15E	3 × GBU-24	1.75	5	53
24 × F-15E	3 × GBU-24	1.75	126	1260
1 × B-2	16 × JDAM[a]	0.33	5	53
16 × B-2	16 × JDAM	0.33	84	840
1 × B-2	16 × JDAM	0.5	8	80
16 × B-2	16 × JDAM	0.5	128	1280

[a]JDAM = Joint Direct Attack Munition.

Again, we point this out not to denigrate the value of the heavy bomber force; indeed, we support its modernization and will argue later in this chapter that the USAF might consider developing a new, long-range strike platform to supplement the existing force. However, enthusiasm for the role bombers can play in power projection must be tempered by the real limitations of their near-term numbers and capabilities.

The second problem with operating mainly from U.S. territory is that for some missions it is simply not a practical option. Consider the complex MOOTW in Burundi described in the previous chapter; the problem there is not putting ordnance on target but supporting complicated and intensive operations on the ground in the heart of Africa. It is difficult to conceive how that could be accomplished in the absence of access to numerous countries in the region, including but not limited to Burundi itself.[11]

We believe that U.S. territory should become an increasingly important launching pad for overseas operations. However, this does not appear to be a complete solution to the access problem.

---

[11]As was pointed out in Chapter Four, even limited operations in Africa have required basing in multiple countries to overcome infrastructure shortfalls.

MANAGING UNCERTAINTY WITH AN ACCESS "PORTFOLIO"

If pure strategies are not adequate to cope with the challenges to come, a hybrid approach is called for. We therefore wish to suggest that the USAF consider a metaphor from the financial world and treat the construction of an appropriate access and basing strategy as a problem in *portfolio management.* We think the analogy is sound along several dimensions:

- As on Wall Street, the environment planners face is one dominated by *uncertainty.* We cannot predict where the next contingency will erupt, what form it will take, or how the geopolitical stars will align to facilitate or restrict the level of international cooperation the United States will receive. In such a "market," a well-hedged portfolio is the best path to success.[12]
- Managing risk and exploiting opportunity require *diversification.* No single investment can ensure maximum financial success; nor can any single-point solution provide a sufficiently robust guarantee of adequate access. Success will depend on having a range of contingency options, plans, and capabilities.
- *Information flows* are critical to good decisionmaking. Just as a competent broker must match the needs of buyers and sellers, so must the United States remain informed and aware of its partners' sometimes-divergent goals, strategies, and interests. *Engagement* and *transparency* play pivotal roles.

What sort of portfolio might the USAF want to construct? In keeping with the metaphor, we will describe one possibility in terms of three components: core investments, hedges against risk, and opportunities to watch out for.

Core Investments

Core investments lie at the heart of our proposed portfolio. They represent secure, low-risk investments that we expect to produce steady results. We will suggest three.

---

[12]On reflection, it is truly unfortunate that the world of international security has no Alan Greenspan it can count on to provide reliable indicators of future circumstances.

The first and most obvious is to recommend that the United States maintain its current array of overseas MOBs in Europe and Asia. These installations are fairly secure and reliable footholds that can serve as points of entry to virtually every region of possible interest. Such bases have in the past been critical to rapid responses to contingencies around the world, and they should continue to play that role into the indefinite future.

Our second recommendation is that the USAF establish a small number of *forward support locations* (FSLs) worldwide. Much discussed under a variety of names, an FSL is essentially a "mega-MOB" intended to support power projection.[13] Spares, equipment, and munitions could be prepositioned at these locations, which should be built where access is either guaranteed or highly likely. FSLs could also host repair facilities for key components such as engines and critical avionics units and would serve as both strategic and intratheater airlift hubs when the situation so demanded.[14] Extensive RAND analysis strongly suggests that properly located and outfitted FSLs offer significant leverage in enabling both rapid and sustainable expeditionary operations.[15]

As Figure 5.1 shows, even a small number of FSLs could provide broad coverage of likely contingency locales. Five FSLs can be found on the figure; in terms of "assured access," three are in U.S. territory (Alaska, Guam, and Puerto Rico), a fourth is on *de facto* U.S. territory (Diego Garcia, at least until 2039), and the fifth is on the territory of America's most reliable ally, Great Britain. Taken together, these five locations put most of the world within C-130 range of a permanent center of U.S. power projection capability.[16] Extreme southern South America and southwestern Africa are left uncovered, but much of the rest of the world's landmass can be served from two different FSLs.

---

[13]RAND has worked extensively on the FSL concept. See, for example, Killingsworth et al. (2000) and Galway et al. (1999).

[14]For a discussion of the kinds of maintenance facilities that might be placed at FSLs, see Peltz et al. (1999).

[15]See Tripp et al. (2000).

[16]These locations also have the virtue of being outside the range of the bulk of any likely adversaries' probable offensive capabilities.

RAND MR1216-5.1

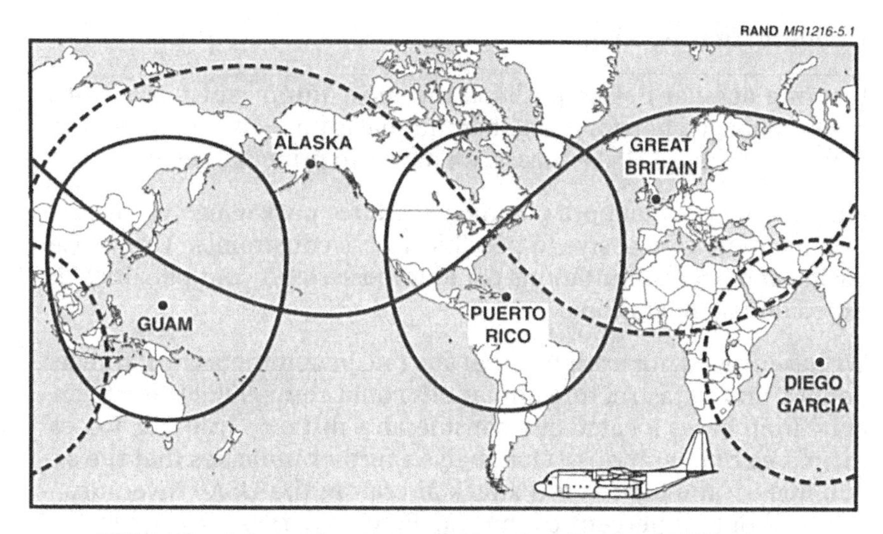

NOTE: The circles on the map are 3000 nm in radius; the AFPAM 10-1403 planning factor for a C-130 is 3200 nm with a 12-ton payload. Solid circles denote an FSL on U.S. territory, while dotted ones are drawn for foreign FSLs.

Figure 5.1—Coverage Available from Five FSLs

Third, the United States should seek to maintain and expand its contacts with key security partners worldwide. Although there would appear to be no need to pursue additional formal defense ties as a means for shoring up prospects for access, consistent engagement is of great value. Training exchanges, joint exercises, and temporary deployments help establish relationships—both formal and, perhaps equally important, informal—that can prove of great value in a crisis. And as U.S. deployments for training and exercises often include engineering undertakings—repairing runways and hardstands, improving fuel storage and delivery facilities, and so forth—they offer opportunities to enhance infrastructure as well as relationships. Finally, these interactions serve to foster the strategic transparency that we believe is invaluable for helping shape partners' perceptions in ways that facilitate future cooperation.

Hedging Against Risk

Insuring against risk is perhaps the most important benefit of a wisely managed portfolio whether the goal is financial or national security. In terms of USAF access, we have two principal suggestions.

First, we suggest that both planning and force packaging may need to become more responsive to possible access constraints. Otherwise, basing and access limitations could impose significant penalties on expeditionary operations.

We have argued, for example, that any one or combination of threats, politics, and infrastructure limitations could compel the USAF to operate from bases located at a considerable distance from the forces' main area of operations. Our analysis further indicates that the capabilities of the fighter and attack aircraft in the USAF inventory—again, about 90 percent of the war-fighting forces—are subject to fairly rapid and dramatic reduction as these distances grow. Prudent steps to offset this decline in effectiveness, such as planning and preparing to provide extra tankers and aircrew to AETFs deploying under such circumstances, are therefore critical hedges that we encourage the Air Force to consider.

The USAF could also consider developing and acquiring some number of high-speed, long-range strike platforms. An aircraft with an unrefueled range of around 2000 nm could, with minimal tanker support, cover most of the world while operating exclusively from the five FSLs we propose, thereby greatly easing the consequences of any future access "lockouts." A cruise speed of around Mach 2 would be valuable in helping the aircraft sustain a reasonable sortie rate.[17] The size of the aircraft—and hence, to first order, its cost—could be kept under control by exploiting the coming generation of small,

[17]The F-15E, for example, has an unrefueled combat radius of approximately 450 to 650 nm depending on loadout, flight profile, and other factors. These aircraft would require at least four refuelings and probably more than 12 hours to complete a 3000-nm radius mission. An aircraft capable of cruising at Mach 2 between refuelings with a 2000-nm range would require 25 percent fewer refuelings and could complete the mission in just under six hours. This would allow the crew of the Mach 2 aircraft to plan and fly a 3000-nm-radius mission every day compared to every other day at best for the crew of a subsonic attack aircraft such as the F-15E.

smart munitions to minimize the weight of its payload while retaining a bomber-like ability to strike multiple targets on each mission.[18]

As an alternative to acquiring a fast, long-range strike aircraft, the USAF could opt to deploy a new generation of long-range munitions for carriage by existing and planned strike aircraft. Current inventory weapons such as the joint standoff weapon (JSOW) and the conventional air-launched cruise missile (CALCM) are relatively few in number and suffer from significant operational limitations.[19] The joint air-to-surface standoff missile (JASSM) promises a range "over 200" nautical miles, but at a fairly high price tag.[20] Further, current plans are for the JASSM to be carried only by heavy bombers and the F-16. Building even more range into weapons employable by more of the current and future fighter force and procuring them in ade-

---

[18]There is a vicious circle to supersonic aircraft design. As the vehicle enters the transonic regime, drag increases immensely and remains high at supersonic speeds owing to the formation of shock waves. Minimizing supersonic drag drives designers to integrate low-diameter engines (turbojets or low-bypass turbofans) rather than the much more fuel efficient high-bypass turbofans. Less efficient engines require additional fuel, which in turn requires additional aircraft structure, which creates extra drag, which demands additional power, and so on. Resolving this cycle of increasing demands for a supersonic bomber has always resulted in a relatively large aircraft for a given payload.

By using munitions such as the 250-lb "small diameter bomb" (SDB), however, the designers of our proposed attack platform could get on the virtuous side of this circle. Trading a 2000-lb weapon for a 250-lb one would actually decrease aircraft gross weight by much more than 1750 lb as the power, fuel, and structure needed to push the big bomb through the "sound barrier" would thereby be shed.

The B-2 carries 16 JDAMs, each weighing a ton and each able to attack a single target. A future "light bomber" could carry weapons weighing an eighth as much apiece so that a total payload of only 3000 to 4000 lb could enable strikes on eight to twelve targets per sortie. Losing those 29,000 lb of payload would mean that a much smaller—and cheaper—aircraft could be built than would otherwise be required. For example, preliminary calculations suggest that a Mach 2 supercruise aircraft with a 2000-nm range and a 2500-lb payload—enough to deliver SDBs onto eight or ten targets per sortie—would weigh in somewhere between 33,000 and 55,000 lb empty weight. This would put it between an F-15 and an F-111 in size.

[19]The JSOW, for example, is a fairly short-ranged (12–40 nm depending on launch altitude) glide bomb. The CALCM has long range and a powerful warhead, but stockpiles are small and it can be launched only from B-52s. JSOW range from the Raytheon Web site, http://www.raytheon.com/es/esproducts/dssjsow/dssjsow.htm, dated August 16, 2000.

[20]JASSMs are expected to cost about $327,000 each; the USAF is planning on buying some 2400. Range figure from Lockheed-Martin, n.d.; cost estimate from U.S. General Accounting Office, 2000; quantity from U.S. General Accounting Office, 1996.

quate numbers could make a major contribution to easing the operational burdens that could arise from access difficulties.

Exploiting Opportunity

A third element of our proposed portfolio is a pair of steps intended to enable the USAF to take advantage of opportunities to significantly improve its access prospects. Investments in these areas are seed money from which little immediate return is necessarily expected but a longer-term bonanza is possible.

We described the first component, "rent-a-rock," earlier. We cannot point to a host country here or there that we believe is primed to lease the United States a chunk of real estate to serve as a military outpost. However, the upside, if such an arrangement could be negotiated (and the construction of the base financed), would potentially be considerable. We suggest that the USAF survey one or more key areas of interest—perhaps starting in the Western Pacific—to see if candidate "rocks" can be identified. If so, some thinking should be done on what kinds of facilities might be called for, how they might be built, and what cost estimates might be developed. Then, it will be prepared should the theoretical possibility of such a deal be transformed into a real opportunity.

Second, the rapid pace of geopolitical change over the past ten years may have created yet-unappreciated opportunities to engage new partners as possible access sites. The countries of Central Asia, for example, have already demonstrated an interest in closer ties with the United States; in case of a crisis involving China or even Iran, Kazakhstan and its neighbors could have great utility as hosts for USAF forces. Similarly, Mongolia, Malaysia, and even Vietnam could help support U.S. actions in Asia, while Israel and the former Soviet republics in the Caucasus could be useful in an SWA contingency.

Areas of Immediate Concern

As a final piece to our portfolio puzzle, we would like to highlight two regions where we believe current access arrangements are insufficient and the risk of being called to action is high. Both immediate and longer-term ameliorative steps may be needed to shore up the USAF position in SWA and much of Asia.

In Southwest Asia, the problem is driven by the seeming impossibility of gaining firm commitments from America's regional friends. We see little prospect of this changing in the immediate future; indeed, as the 1991 Gulf War fades into ever more distant memory, pressures may begin to grow in Saudi Arabia and elsewhere to impose further limits on U.S. presence and access on the Arabian peninsula.

In the near term, we believe that flexible planning will be critical to ensuring the USAF's ability to effectively fly and fight in the Persian Gulf. Enabling deploying forces to maintain OPTEMPO from nonoptimal basing locations could be vitally important in this region. Looking out further, broadening the list of possible strategic partners is advisable, with Israel being a prime candidate should a broad peace accord permit its "normalization" in the region.

The Pacific Rim, meanwhile, offers increasing challenges the further south one casts one's eyes.[21] The current USAF basing posture is wholly inadequate to support high-intensity combat operations anywhere much beyond the Korean peninsula. Especially problematic is the lack of bases available in the vicinity of the Taiwan Strait. Renewed access to bases in the northern Philippines could be immensely helpful here, especially if confidence were high that these bases could be used if a fight erupted between the mainland and Taiwan. Such political concerns—which are rife with regard to Taiwan throughout the region—would make "rent-a-rock" a particularly attractive option here.

Still further south, the United States may want to consider taking steps to improve its access prospects by increasing the level and extent of its presence in Singapore. The United States should seek to build further on its excellent relations with Thailand and continue to assess Malaysia as an option for the future, depending in large part on future political developments. Vietnam may also be a longer-term alternative.

---

[21]A discussion of some Asian basing issues can be found in Khalilzad et al. (2001).

In addition, an increased number of longer-range combat platforms (or short-legged platforms with long-range munitions) would be useful in both the Gulf and East Asia.

CONCLUDING REMARKS

There may come a time when many of the access issues we have discussed are no longer of concern. One can imagine, for example, a future in which space-based surveillance and strike systems enable responsive strikes on any target, moving or stationary, anywhere in the world.[22] In that tomorrow, the need to deploy combat aircraft—and thousands of airmen and airwomen—to distant shores to fight their nation's wars will have ended, and much of what is covered in this report will be of little more than historical interest. However, even come that day of jubilee, we are willing to wager that it will still be difficult to get food, water, and medical care to threatened people in Rwanda or Tierra del Fuego, or to separate warring factions in the Balkans or East Timor. And so long as nations continue to jealously exercise control over their land, air, and water, the Air Force will from time to time come up against difficulties relating to access and basing.

Our research indicates that there is no panacea or "silver bullet" awaiting discovery. Old problems, like the vagaries of international politics, will persist, and new ones—dozens or even hundreds of long-range, accurate missiles aimed at U.S. bases—will emerge. Furthermore, nothing comes free: There are real costs, in terms of both money and opportunity, associated with any course of action the USAF might take to deal with potential problems in this area. This is the bad news.

On the other hand, we do not emerge from our work with nothing but a tale of woe. We believe that the problems we have discussed are manageable and that even those that can't be foreseen—always the most worrisome—can be minimized through a well-thought-out global access strategy. The strategy we suggest calls for increased flexibility and pays off in enhanced robustness against the in-

---

[22]Please see Preston et al. (2002).

eluctable uncertainty that characterizes this problem. In the final analysis, then, access is not a problem to be solved—it is a portfolio to be managed.

Appendix

AIRCRAFT CONFIGURATIONS AND RAMP REQUIREMENTS

This appendix provides some additional detail on the aircraft configurations and parking-space calculations used in Chapter Three.

AIRCRAFT CONFIGURATIONS

Tables A.1 through A.4 depict the detailed aircraft configurations used to determine weight and drag profiles.

RAMP SPACE REQUIREMENTS

Fighter parking space requirements were based on AFH 32-1084, Table 2.6. For example, each F-15C requires a block 54 × 75 feet, or 4050 square feet. Eighteen aircraft require 4050 × 18 = 72,900 square feet. Assuming the aircraft are parked in two lines of nine aircraft facing each other, the taxiway between them must be 90 feet wide and 675 feet long for a total of 60,750 square feet. This gives a total required ramp space of 133,650 square feet for the 18 F-15Cs. Space requirements for the other fighters were computed using the same method. Figure A.1 shows how the 360,000-square-foot area is divided among the various fighters of a typical AETF. Similar calculations were done for the support aircraft, with the results shown in Figure A.2.

Table A.1

A-10 Configurations

Item	Weight (lb)	Fuel (lb)	Drag Index
A-10	28,000	0	0
Internal fuel	10,700	10,700	0
Chaff/flares	328	0	0
1 × DRA + 2 LAU-105	161	0	0.23
2 × AIM-9	382	0	0.40
1 × ALQ-184-7	631	0	0.99
2 × LAU-88/A	930	0	1.00
6 × AGM-65G	3,990	0	4.92
Drag due to asymmetric load	0	0	0.10
Total	45,122	10,700	7.24
1 × 600-gallon tank	4,403	3,961	1.80
Total	49,525	14,661	9.44

Table A.2

F-15C Configurations

Item	Weight (lb)	Fuel (lb)	Drag Index
F-15C	29,500	0	0
Internal fuel	13,500	13,500	0
20-mm ammo	531	0	0
4 × AIM-120	1552	0	5.2
4 × AIM-9L/M	780	0	8.4
4 × LAU-128/A	444	0	4.8
3 × 610-gallon tanks	12,855	11,895	33.1
Total	59,162	25,395	51.5

Table A.3

F-15E Configurations

Item	Weight (lb)	Fuel (lb)	Drag Index
F-15E	34,600	0	0
Internal fuel	12,915	12,915	0
2 × CFT	13,738	9,352	21.3
20-mm ammo	289	0	0
2 × AIM-120	676	0	3.4
2 × AIM-9L/M	390	0	4.2
2 × LAU-128/A	222	0	2.2
2 × 610-gallon tanks	8,570	7,930	24.6
LANTIRN pods	1,141	0	16.9
Total	72,541	30,197	72.6
2 × GBU-12	1,220	0	8.6
Total	73,761	30,197	81.2
4 × GBU-12	2,440	0	17.2
Total	74,981	30,197	89.8
4 × GBU-24	9,292	0	24.8
Total	81,833	30,197	97.4

Table A.4

F-16C Configurations

Item	Weight (lb)	Fuel (lb)	Drag Index
F-16C	18,700	0	0
Internal fuel	7,162	7,162	0
20-mm ammo	287	0	0
Chaff/flares	130	0	0
2 × AIM-120	682	0	0
2 × 370-gallon tanks	5,982	4,800	35
1 × ALQ-184-5	471	0	18
Pylon/adapter	217	0	11
Total	33,631	11,962	64
1 × LANTIRN pod	429	0	32
2 × GBU-12	1,222	0	14
2 × TER[a]	818	0	34
Total	36,100	11,962	144
1 × LANTIRN pod	429	0	32
2 × GBU-24	4,708	0	40
Total	39,586	11,962	136
2 × CBU-87	1,900	0	36
Total	36,349	11,962	100

[a]Triple-ejector rack.

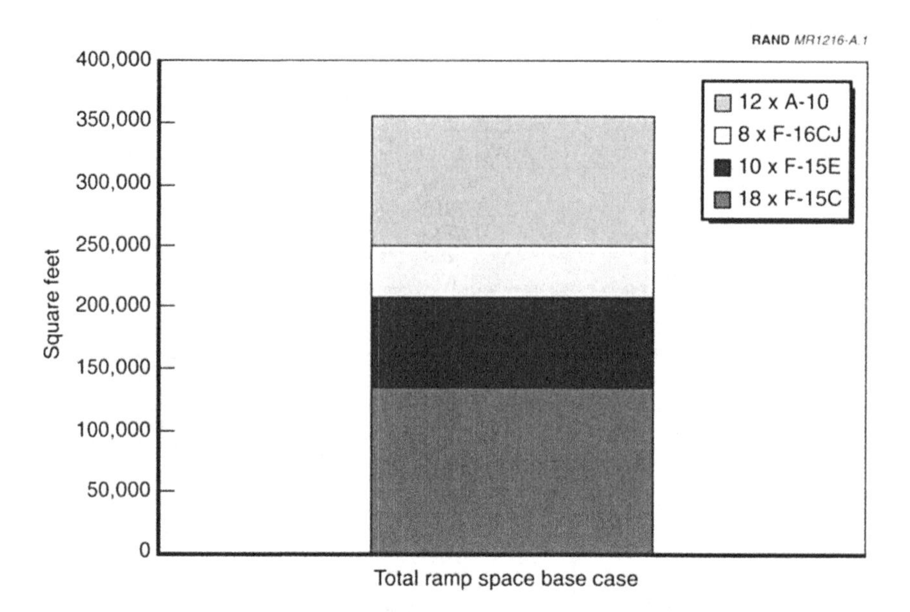

Figure A.1— Fighter Parking Ramp Space Required

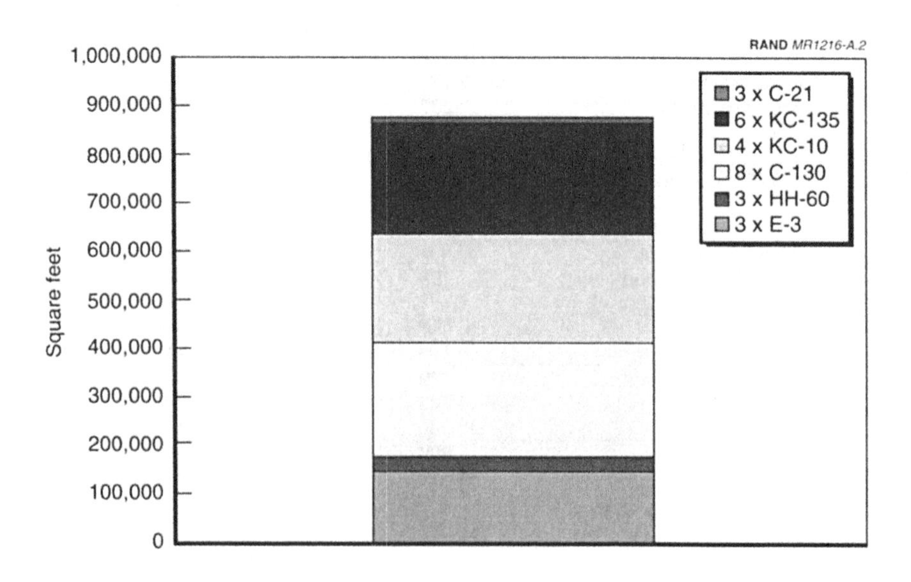

Figure A.2—Support Aircraft Parking Ramp Space Required

BIBLIOGRAPHY

Abdallah, Samih, "Interview with Greek Foreign Minister George Papandreou," *al-Akhbar* (Cairo), May 29, 1999, p. 7, translated in "Daily Interviews Greek Foreign Minister," *FBIS Daily Report*, May 29, 1999.

Abel, David, "Holes Open in U.S. Drug-Fighting Net," *Christian Science Monitor*, July 28, 1999.

"The Access Issue," *Air Force Magazine*, October 1998.

Aguirre, Mariano, "Spain's 'Nuclear Allergy': The U.S. Finds a Treatment," *The Nation*, Vol. 247, No. 20, December 26, 1988.

"Air Ban on Turkish Fighter Planes, *The Herald* (Glasgow), May 22, 1999, p. 10.

Albor, Teresa, "Volcano Damage, Nuclear Issue Complicate Philippine Base Talks," *Christian Science Monitor*, June 26, 1991.

Allard, Kenneth, *Somalia Operations: Lessons Learned*, National Defense University Press, Fort McNair, Washington, D.C., January 1995.

"Allies Wanted All-out Attack on Khadafy, Reagan Says," *Toronto Star*, April 22, 1986, p. A1.

"Armed Forces to Join Drug Enforcement Effort," *O Globo* (Rio de Janeiro), April 17, 1996, translated in *FBIS Daily Report*, April 17, 1996.

"Baku Asks for US Support," *Financial Times*, June 19, 1999, p. 4.

"Balkan States Back NATO, Fear Instability," *Agence France Presse*, April 26, 1999.

Blaustein, Susan, "Steamrollered One More Time: U.S. Military Bases in the Philippines," *The Nation*, Vol. 253, No. 7, September 9, 1991.

Bowden, Mark, *Black Hawk Down*, New York: Atlantic Monthly Press, 1999.

Boyne, Walter J., "Nickel Grass," *Air Force Magazine*, December 1998, pp. 54–59.

_______, "El Dorado Canyon," *Air Force Magazine*, March 1999, pp. 56–62.

"Brave Gamble," *The Economist*, May 29, 1999.

Briscoe, David, "U.S. Officials Cite Possible Removal of Bases from Philippines," *Associated Press*, September 16, 1988.

Bruce, James, "Saudi/U.S. Relations Strained by Policies Towards Iraq and Israel," *Jane's Defense Weekly*, September 25, 1996, p. 29.

Builder, Carl H., and T. W. Karasik, *Organizing, Training and Equipping the Air Force for Crises and Lesser Conflicts*, Santa Monica: RAND, MR-626-AF, 1995.

Chacon, Ronald Moya, "Ministers Deny Report on Transfer of U.S. Bases," *La Nacion* (San Jose), February 13, 1999, translated in *FBIS Daily Report*, February 13, 1999.

Christian, Derek J., "The African Crisis Response Force: A Critical Issue for Africa," *Naval War College Review*, No. 3, Summer 1998, pp. 70–81.

Church, George J., "Hitting the Source," *Time*, April 28, 1986.

Cody, Edward, "U.S. and Spain Face Showdown on Bases," *Washington Post*, November 6, 1987, p. A23.

Cohen, Elliot, et al., *Gulf War Air Power Survey*, Vol. V, *Statistical Compendium and Chronology*, Washington, D.C.: U.S. Government Printing Office, 1993.

Comptroller General of the United States, *Report to the Congress: Airlift Operations of the Military Airlift Command During the 1973 Middle East War*, U.S. General Accounting Office, LCD-75-204, 1975.

Cook, Donald, *Evolving to an Expeditionary Aerospace Force: The Next Air Force*, USAF, Directorate for EAF Implementation, September 1998.

Cushman, John H., Jr., "Danger in the Gulf: How U.S. Navy Girds for Escort Duties With Tankers," *New York Times*, July 19, 1987, p. 1.

_______, "Navy to End Convoys in Gulf But It Will Protect Ships," *New York Times*, September 17, 1988, p. 2.

Doerner, William R., "In the Dead of Night," *Time*, April 28, 1986.

Farah, Douglas, "Handover of Panama Base Hinders Anti-Drug Efforts," *Washington Post*, May 30, 1999, p. A19.

Feil, Scott, *Preventing Genocide: How the Early Use of Force Might Have Succeeded in Rwanda*, New York: Carnegie Foundation, 1998.

"The First 8 Days: Chronicle of the Development of Combat Operations," *Nezavisimoye Voyennoye Obozreniye* (Moscow), No. 12, April 2, 1999, p. 2, translated in "The First 8 Days of NATO Airstrikes," *FBIS Daily Report*, April 2, 1999.

Fitchett, Joseph, "NATO Summit Charts a Kosovo Policy," *International Herald Tribune*, April 26, 1999, p. 1.

"Foreign Minister Zulfugarov Says NATO Bases in Azerbaijan Possible," *BBC Summary of World Broadcasts*, August 27, 1999.

Fulghum, David A., "USAF Plans Rapid, All-Stealth Task Force," *Aviation Week & Space Technology*, February 26, 2001, p. 24.

Fulghum, David A., and Robert Wall, "U.S. To Move First, Plan Details Later," *Aviation Week & Space Technology*, September 24, 2001, p. 40.

Galway, Lionel, et al., "Expeditionary Airpower: A Global Infrastructure to Support EAF," *Air Force Journal of Logistics*, Vol. 23, No. 2, Summer 1999.

Gedda, George, "Fearing China, Manila Turns to U.S.," *Fort Worth Star-Telegram*, July 6, 1999.

Gourevitch, Philip, *We Wish to Inform You That Tomorrow We Will Be Killed With Our Families*, New York: Farrar, Straus and Giroux, 1998.

Grossman, Elaine M., "Commander Says Pact for Base Access in Ecuador Is Close at Hand," *Inside the Pentagon*, September 23, 1999.

Hawley, Richard E., et al., "Global Reconnaissance-Strike: Innovative Concept Leverages Existing Programs for Early Answer to Anti-Access Challenge," *Armed Forces Journal International*, June 2000, pp. 52–57.

Hersh, Seymour M., "Target Qaddafi," *New York Times*, February 22, 1987, Section 6.

Heyman, Charles (ed.), *Jane's World Armies*, Jane's Information Group, 1999.

Hirsh, John L., and R. Oakley, *Somalia and Operation Restore Hope: Reflections on Peacemaking and Peacekeeping*, Washington, D.C.: United States Institute of Peace Press, 1996.

Hua, Lee Siew, "Region 'More Receptive' to U.S. Military Presence," *The Straits Times* (Singapore), November 10, 1998, p. 1.

Jehl, Douglas, "Attack on Iraq: In the Gulf," *New York Times*, December 19, 1998, p. A9.

_______, "Saudis Restrict Use of U.S. Aircraft Against Iraq," *New York Times*, March 22, 1999.

Jelinek, Pauline, "Military Looks to Cut Patrols in U.S.," *Los Angeles Times*, January 14, 2002.

Jordan, Michael J., "NATO Enlists a Reluctant Hungary Into Kosovo War," *Christian Science Monitor*, June 2, 1999a, p. 6.

_______, "NATO Wants More than Words from Hungary," *Radio Free Europe/Radio Liberty Newsline*, Vol. 3, No. 111, Part II, June 8, 1999b.

Khalilzad, Zalmay, David Shlapak, and Daniel Byman, *The Implications of the Possible End of the Arab-Israeli Conflict for Gulf Security*, Santa Monica: RAND, MR-822-AF, 1997.

Khalilzad, Zalmay, David T. Orletsky, Jonathan D. Pollack, Kevin Pollpeter, Angel M. Rabasa, David A. Shlapak, Abram N. Shulsky, and Ashley J. Tellis, *The United States and Asia: Toward a New U.S. Strategy and Force Posture*, RAND, MR-1315-AF, 2001.

Killingsworth, Paul S., Lionel Galway, Eiichi Kamiya, Brian Nichiporuk, Timothy L. Ramey, Robert S. Tripp, and James C. Wendt, *Flexbasing: Achieving Global Presence for Expeditionary Aerospace Forces*, RAND, MR-1113-AF, 2000.

Kozaryn, Linda D., "Military Rep Builds Bonds in South Africa," Armed Forces Information Service News Articles, http://www.defenselink.mil/ news/May 1999/n05261999_9905264.html, updated May 26, 1999.

Kuperman, Alan J., "Rwanda in Retrospect," *Foreign Affairs*, Vol. 79, No. 1, January/February 2000, pp. 94–118.

Lichfield, John, "UK's Base in Indian Ocean Used for Strike," *Independent*, September 4, 1996, p. 8.

Lockheed-Martin Corporation, Joint Air-to-Surface Standoff Missile Fact Sheet, 2002. Available at http://www.jassm.com/about.html.

Lund, John, "The Airlift to Israel Revisited," unpublished manuscript, 1990.

"Manila Says Subic Naval Base Will Be Closed by End of 1992," *New York Times*, December 27, 1991, p. A6.

Mann, Jim, "U.S. Facing Crisis Over Global Military Bases," *Los Angeles Times*, May 15, 1988, p. 1.

Military Traffic Management Command, Transportation Engineering Agency, *Deployment Planning Guide*, MTMCTEA Ref. 97-700-5, July 1997.

Mitchell, Randy, and Kenneth Fidler, "Food Airdrop to Afghans Underscores President's Humanitarian Pledge," *Air Force Link* Web site, http://www.af.mil/news/n20011008_1422.shtml, October 8, 2001.

"NATO Deployed 1,100 Warplanes Against FRY in Campaign," *AFP* (North European Service-Paris), June 10, 1999, *FBIS Daily Report*, June 10, 1999.

"NATO Launches Bombing Raids from Hungary," *AFP* (North European Service-Paris), May 28, 1999, *FBIS Daily Report*, May 28, 1999.

Olonisakin, Funmi, "African Homemade Peacekeeping Initiatives," *Armed Forces and Society*, Vol. 23, No. 3, Spring 1997, pp. 349–372.

"Orthodox but Unorthodox," *The Economist*, April 17, 1999.

Owen, Ivor, and Kevin Brown, "Ex-Prime Ministers Oppose Thatcher Decision Over Air Bases," *Financial Times*, April 17, 1986, p. 10.

Peltz, Eric, Hyman L. Shulman, Robert S. Tripp, Timothy Ramey, and John G. Drew, *Supporting Expeditionary Aerospace Forces: An Analysis of F-15 Avionics Options*, Santa Monica: RAND, MR-1174-AF, 2000.

"Philippine Senate Rejects U.S. Base Deal; Aquino Planning to Take Subic Bay Issue to People," *Seattle Post-Intelligencer*, September 16, 1991, p. A1.

Pirnie, Bruce, *Civilians and Soldiers: Achieving Better Coordination*, Santa Monica: RAND, MR-1026-SRF, 1998.

Pirnie, Bruce, and C. Francisco, *Assessing Requirements for Peacekeeping, Humanitarian Assistance, and Disaster Relief*, Santa Monica: RAND, MR-951-OSD, 1998.

Preston, Bob, Dana J. Johnson, Sean Edwards, Jennifer Gross, Michael Miller, and Calvin Shipbaugh, *Space Weapons, Earth Wars*, Santa Monica: RAND, MR-1209-AF, 2002.

Riding, Alan, "Torrejon Journal; The F-16 Drama (Cont.): Happy Landing in Italy?" *New York Times*, February 6, 1990, p. A4.

"Saudis Not to Let U.S. Launch Attack from Territory," *AFP* (North European Service—Paris), November 14, 1998, *FBIS Daily Report*, November 14, 1998.

Schmitt, Eric, "Pentagon Worries About Cost of Aid Missions," *New York Times*, August 5, 1994, p. A6.

Schumacher, Edward, "U.S.-Spanish Discord Over Bases Is Growing," *New York Times*, December 14, 1986, p 6.

Schutz, Barry, "There's Something About Africa: U.S. National Security Interests in Sub-Saharan Africa," *National Security Studies Quarterly*, Autumn 1998.

Sciolino, Elaine, "U.S. and Philippines Sign Pact on Bases," *New York Times*, October 18, 1988, p. A3.

Sherbrooke, Craig, *Using Sorties vs. Flying Hours to Predict Aircraft Spares Demand*, McLean, VA: Logistics Management Institute, April 1997.

Sisk, Richard, "Regional Allies Hinder U.S. Efforts," *New York Daily News*, September 4, 1996, p. 2.

Siverson, Randolph M., and Joel King, "Attributes of National Alliance Membership and War Participation, 1815–1965," *American Journal of Political Science*, Vol. 24, No. 1, February 1980.

Sly, Liz, "A Welcome Mat in the Balkans," *Chicago Tribune*, May 11, 1999, p. 6.

Smith, Alastair, "To Intervene or Not to Intervene," *Journal of Conflict Resolution*, Vol. 40, No. 1, March 1996.

Stanik, Joseph T., "Welcome to El Dorado Canyon," *Proceedings of the United States Naval Institute*, Vol. 122, April 1996, pp. 57–62.

Steele, Jonathan, "Why Spain Is Talking Tough on US Bases," *Manchester Guardian Weekly*, April 26, 1987.

"Stifling U.S. Pressure," *Ta Nea*, May 28, 1999, p. 18, translated in "United States Said 'Dissatisfied' with Greece on Kosovo," *FBIS Daily Report*, May 28, 1999.

Stillion, John, and D. T. Orletsky, *Airbase Vulnerability to Conventional Cruise-Missile and Ballistic-Missile Attacks: Technology, Scenarios, and U.S. Air Force Responses*, Santa Monica: RAND, MR-1028-AF, 1999.

Storey, Ian, "Manila Looks to USA for Help Over Spratlys," *Jane's Intelligence Review*, August 1, 1999.

Suarez, Miguel C., "A Week After Breaking Down, Bases Talks Back on Track," Associated Press, August 2, 1988.

Szamado, Eszter, "NATO Launches Bombing Raids from Hungary," Paris AFP (North European Service) in English 1608 GMT May 28, 1999, from *FBIS Daily Report (*East Europe), May 28, 1999.

Tagliabue, John, "Crisis in the Balkans: Repercussions," *New York Times*, May 12, 1999, p. A14.

Timsar, Richard, "Lisbon Woos Arabs, Closes Azores to Future Airlift of U.S. Arms to Israel," *Christian Science Monitor*, April 27, 1981, p. 13.

Tirpak, John A., "The Indispensable Fighter," *Air Force*, March 2001, pp. 22–29.

Tripp, Robert S., L. A. Galway, T. L. Ramey, M. Amouzegar, and E. Peltz, *Supporting Expeditionary Aerospace Forces: A Concept for Evolving to the Agile Combat Support/Mobility System of the Future*, Santa Monica: RAND, MR-1179-AF, 2000.

Tyler, Patrick E., "Carlucci Sees Limited U.S. Gulf Role: Expansion of Naval Escort Mission Is Called Unlikely," *Washington Post*, January 6, 1988a, p. A17.

_______, "U.S. Gulf Force to Scale Back Escort Duty," *Washington Post*, December 8, 1988b, p. A39.

U.S. Air Force Handbook, AFH 32-1084, Chapter 2.

U.S. Air Force Air Mobility Command, *Air Mobility Planning Factors*, AFPAM 10-1403, June 1997.

"U.S. Defense Secretary, Ukrainian Leaders Discuss New Military Cooperation Programs," *Jamestown Foundation Monitor*, Vol. 5, No. 148, August 2, 1999.

U.S. Department of Defense, "The United States Security Strategy for the East Asia-Pacific Region 1998," available at http://www.defenselink.mil/pubs/easr98/.

U.S. General Accounting Office, *Defense Acquisitions: Need to Revise Acquisition Strategy to Reduce Risk for Joint Air-to-Surface Standoff Missile*, Washington, D.C.: U.S. Government Printing Office, GAO/NSIAD-00-75, April 2000.

U.S. General Accounting Office, *Precision-Guided Munitions: Acquisition Plans for the Joint Air-to-Surface Standoff Missile*, Washington, D.C.: U.S. Government Printing Office, GAO/NSIAD-96-144, June 1996.

"U.S., Spain Announce Withdrawal of U.S. F-16s," Associated Press, January 15, 1988.

Wolfe, Frank, "Jumper Lays Out Future CONOPs for Global Strike Task Force," *Defense Daily*, February 20, 2001, p. 5.

Woodward, Bob, *The Commanders*, New York: Simon and Schuster, 1991.

Wright, Robin, and William D. Montalbano, "U.S. Launches Missile Attack on Iraq Targets," *Los Angeles Times*, September 4, 1996, p. A1.

Made in the USA
Middletown, DE
17 November 2023